DESIGN in

WHAT HAPPENS BEHIND THE

PROGRESS

To My Family

Thanks to all the designers who spared their time and artwork to make this endeavor a success. And to Dave, whose patience and talents were an incredible asset in the design of this book. To the rest of the staff at Supon Design Group—Andy, Dianne, Rick, Tony, and Wayne—who helped realize this book's smooth completion. To Bradley and Linda for their writing. And finally, to Henry Kornman of Books Nippan, who saw an opportunity for this book and made it possible.

— *Supon Phornirunlit*

DISTRIBUTED TO THE TRADE IN THE UNITED STATES, CANADA AND MEXICO BY:

1123 Dominguez Street, Unit K
Carson, CA 90746
FAX: (310) 604-1134

DISTRIBUTED THROUGHOUT THE REST OF THE WORLD BY:
Nippon Shuppan Hanbai Inc.
3-4 Chome, Kandasurugadai,
Chiyoda-ku
Tokyo 101, Japan
FAX: (03) 3233-1578

PUBLISHER:

1123 Dominguez Street, Unit K
Carson, CA 90746

DESIGN IN PROGRESS IS A PROJECT OF:

Supon Design Group,
International Book Division
1000 Connecticut Ave, NW
Suite 415
Washington, DC 20036
Tel: (202) 822-6540

Printed in Hong Kong

ISBN 0-945814-05-4

Library of Congress Catalog Card Number 92-64014

DESIGN in

WHAT HAPPENS BEHIND THE SCENES

PROGRESS

SU N DESIGN ROUP

ACKNOWLEDGMENTS

ART DIRECTOR AND EDITOR
SUPON PHORNIRUNLIT

MANAGING EDITOR
WAYNE KURIE

WRITERS
LINDA KLINGER
BRADLEY RYMPH

DESIGN EDITOR
SUPON PHORNIRUNLIT

DESIGNER
DAVE PRESCOTT

SUPPORTING STAFF
DIANNE COOK
ANDREW DOLAN
RICK HEFFNER
TONY WILKERSON

DESKTOP PUBLISHING
microPRINT,
WASHINGTON, DC

CAMERA SERVICES
QUADRANT SLIDE SERVICE

TABLE OF CONTENTS

INTRODUCTION

Years ago, M. H. Alderson confirmed it: If at first you don't succeed, you're running about average. That observation especially rings true in the design world, where seldom does one create the perfect piece with a single, swift pencil sketch.

Part of the enjoyment and the magic of graphic art is the process of piecing together the puzzle: What if I put this line there? that color here? use this type style in the foreground and set it at an angle on the page? Like amateur chemists, we experiment with light and perception, imagination and attitude. Our attempts illuminate, ignite, sometimes smolder—but they all become part of the invisible strand connecting origin to the final product. Upon our cast-offs, excellence is solidly built.

Award-winning works undergo many solutions on the road to recognition. Even the best graphic designer spends much time envisioning approaches for a project, then carefully giving form to his or her visions through sketches and compositions—only to have all the ideas later rejected.

Design in Progress: What Happens Behind the Scenes is a celebration of design creation. The 53 projects it

includes—from the late 1980s and early 1990s—represent both the finished piece and those stages that preceded it. The 22 design firms whose works are illustrated in these pages are based in Australia, Canada, European Countries, and Japan, as well as large and small cities across the United States. They include some of the world's most famous studios and smaller studios with primarily local clienteles and reputations. All share an important characteristic: They know that quality design is not simply a matter of artistic talent. Rather, it requires a careful blending of creativity, flexibility, and diplomacy, enabling designers to work with clients to explore what makes each project unique in its needs and its opportunities.

Obviously, a concept is not initiated with the completed product, but oftentimes that product appears so appropriate and so exact that audiences have no idea of the multiple iterations it has already weathered, and the hidden hours devoted to conceptualizing, imagining, and creating. They don't see the multitude of comps produced and discarded, the customizing that eventually adapted the piece to mirror the client's image in his own eye.

As designers, we're commissioned by a wide range of clients to create a variety of visual works, from corporate logos and identity campaigns to book covers and promotional posters. We've learned not to equate "simple design" with an easy task—often the job using only a few graphic elements to communicate a powerful message is the job that took an extraordinary amount of effort. Nor is the route to excellence always direct; sometimes it's circuitous, bringing you back to where you started. Similar projects may resist flowing in the same direction or may not use the same procedures to travel from start to finish. What we think about along the way tells us a lot about ourselves and about the process of creativity. It also teaches us how successful design is borne out of a rite of trial and error.

This book contains numerous examples of this learning experience. Whether our task is to present several comps at once, or to condense numerous factors and develop just one idea for client approval, examining the method used can be a very rewarding pastime. Sometimes the process starts with a direct request from the client: Capture the

low-lit, intimate atmosphere of this pub, or recast this corporation's position in the international marketplace. With one of the projects included here, a studio was given the opportunity to begin by coming up with a name for a new restaurant; after the studio created a list with great visual potential and the client made his selection, it was able to develop a complete identity package—signage, menus, T-shirts, bags, etc.—giving graphic life to that name.

Another project required a studio to simultaneously explore a variety of logo approaches as a newly formed corporation wrestled with what to call itself.

Many design projects, however, modify an existing business name or product to incorporate a new direction, an acquisition, a proud heritage. Designers update different projects in a variety of ways—by using new photographs or illustrations for print materials, a unique lid, handle or shape for packaging, or unusual paper and binding for a calendar. And regardless of where it starts, the client is always an integral part of the final result. One studio was hired by a small international airline to do a complete analysis of the travel market relating to its business. This entailed the designers' helping the airline expand its competitiveness by overhauling its entire visual identity—logo, printed materials, aircraft graphic design, staff uniforms. And another studio was asked to develop entirely new and elaborate packaging for its luxury-market liqueur.

This book explores the designer's thoughts and the succession of artwork created along the way, as graphic elements are acquired and omitted, and guidelines emerge that address the client's intentions with a fresh look. Here, you'll find an interesting collection of comprehensive art forgotten or filed away after the project is completed. Observe the changes in color, line, typography, and dimension from one stage to the next. One may start out with a playful image and redirect the piece to a more somber tone, one that communicates far more by understatement. Or type may be used to spell out a message, but metamorphises into a complex illustrative identity that wins awards the world over. These rejected presentations provide a vital educational link between concept and task completion.

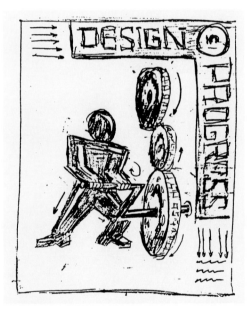

Designing is an evolution. To many, it is an especially pleasant journey. The end result, of course, is what wins recognition, but the process is what gives the result meaning. As Eugene Delacroix said, "We work not only to produce, but to give value to time." *Design in Progress* showcases the invaluable effort designers exert creating lasting impressions.

Owner and art director of the Washington, D.C.-based Supon Design Group, Inc., Supon was born in Bangkok and studied design in Thailand, and the United States. Currently, Supon is on the board of directors of the Art Directors Club of Metropolitan Washington and is project director of the Washington Trademark Design bi-annual competitions. Since opening his studio in 1988, Supon has been featured in several publications, including the Washington Journal of AIGA, How's 1990 Business Annual, Self Image, *and* Asia's Media Delite. *In the past four years, Supon Design Group has earned over 200 awards from the Art Directors Clubs of both New York and Washington, AIGA, Type Directors Club,* American Corporate Identity, *DESI,* Print's Design Annual, *and many others. The studio has authored several books exploring the nuances of graphic design, and its work has been exhibited in England, Israel, Japan, Thailand and the United States.*

PROJECTS

ACTIVE8 — BROCHURE

DESIGN FIRM
GRAFFITO

ART DIRECTORS
TIM THOMPSON,
MORTON JACKSON

DESIGNERS
TIM THOMPSON, MORTON
JACKSON, JOE PARISI

PHOTOGRAPHERS
TARAN Z, HOWARD
EHRENFELD, ED WHITMAN,
MORTON JACKSON

WRITER
TONY MAFALE

CLIENT
ACTIVE8

Active8 needed promotional material conveying its position on the cutting edge of interactive communications technology. Graffito, an East Coast design studio that was one of the partners in the newly established company, was given the responsibility of producing this capabilities brochure.

After Graffito finished the copy and laid out the brochure spread by spread, it used its computer system to scan in the various photos and finalize the comp.

The printed product was given a spiral binding so that selected sections could be replaced without reprinting the whole booklet.

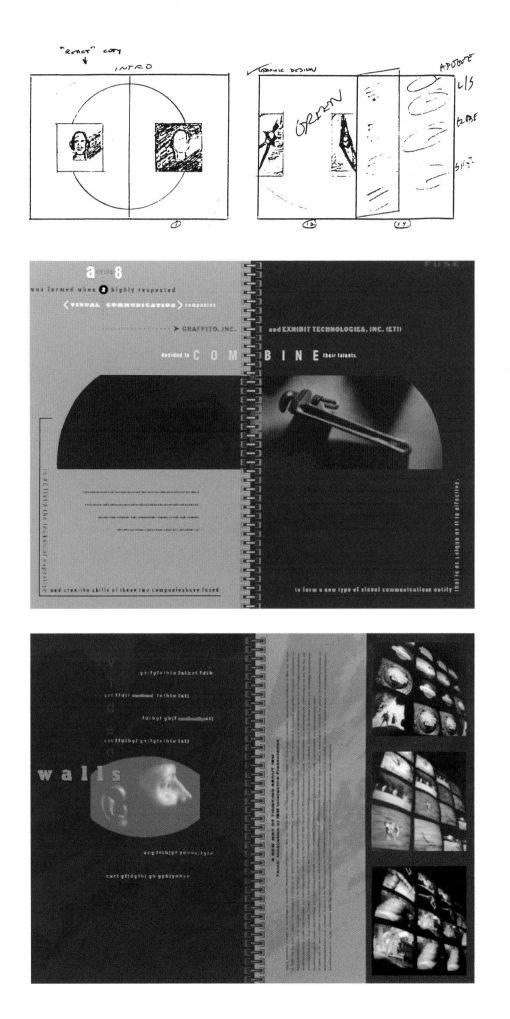

AGENDA – REPORT

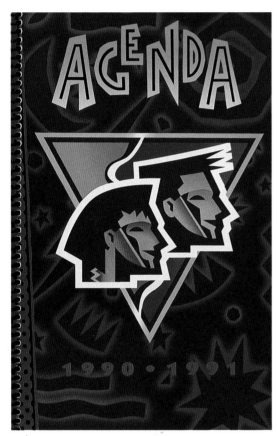

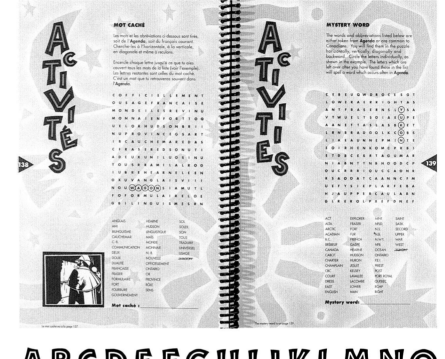

DESIGN FIRM
TURQUOISE DESIGN

ART DIRECTOR
MARK TIMMINGS

DESIGNER
MARIO GODBOUT

ILLUSTRATOR
JEAN SOULARD

CLIENT
OFFICE OF THE
COMMISSIONER OF
OFFICIAL LANGUAGES

The Office of the Commissioner of Official Languages annually produces an *Agenda* to teach Grade 11 students why English and French are both official languages in Canada. Turquoise Design began creating *Agenda 1990–91* by researching the visual styles popular among teens, then created images suggestive of musical videos. An original typeface was developed to complement the illustrations. Both the history of Canadian bilingualism and the subject of youth were treated with an attitude of humor and respect—an approach that helped make students more open to the subject matter.

The Office of the Commissioner of Official Langues is pleased to present you with your copy of **Agenda** 90-91. In it you'll find interesting information about our two official languages and about both English-speaking Canadians who have left their mark on our history, from the first European explorers down to the present day. Short exercises are also included to enable you to test your knowledge in this area.

You will find that English and French have been used in Canada for a long time. *The Official Languages Act*, adopted in 1988, confirms the equal status of two official languages and strengthens language rights. It also recognizes the importance of our many other languages.

The Commissioner of Official Languages is Canada's language ombudsman. If you have trouble getting services in the official language of your choice in a federal institution, you can write or call the Commissioner's office nearest you. The address and telephone numbers are found on the last page of **Agenda**.

Agenda 1990-1991 is designed to help you plan and organize your school and personal activities. The first section provides some tips on how to go about it. You can use **Agenda** to record important things to do, course schedules, examinations and appointments as well as friends' birthdays, addresses and telephone numbers. Make use of it daily.

Do you like this year's look? Do you find **Agenda** useful and interesting? Send us your impressions, comments or suggestions. You can send them to the Office of the Commissioner of Official Languages in Ottawa. Have a good school year!

C'est avec plaisir que le Commissariat aux langues officielles t'offre l'édition 1990-1991 de son **Agenda**. Tu y trouveras toutes sortes de renseignements sur les deux langues officielles et sur les Canadiens d'expression française et anglaise qui ont marqué notre histoire, des premiers explorateurs européens à nos jours. De courts exercices sont même inclus pour te permettre de vérifier tes connaissances dans ce domaine.

Tu verras que la présence du français et de l'anglais au Canada remonte à bien longtemps. Quant à la *Loi sur les langues officielles* promulguée en 1988, tout en reconnaissant l'importance de nos diverses autres langues, elle confirme le statut égal de nos deux langues officielles et renforce les droits qui s'y rattachent.

Le Commissaire aux langues officielles est l'ombudsman linguistique des Canadiens. Si tu as de la difficulté à te faire servir dans la langue officielle de ton choix par une institution fédérale, tu peux écrire ou téléphoner au bureau du Commissaire le plus proche. Les adresses et numéros de téléphone se trouvent à la dernière page de l'**Agenda**.

L'**Agenda** 1990-1991 est conçu pour t'aider à planifier et à organiser tes activités scolaires et personnelles. La première section te propose quelques trucs pour y arriver. Tu pourras noter dans ton agenda une foule de renseignements : tes priorités, tes heures de cours, d'examens et de rendez-vous, des remarques, les anniversaires, les adresses et numéros de téléphone de tes amis. Fais-en ton compagnon de tous les jours.

La présentation graphique de cette année te plaît-elle ? Trouves-tu cet agenda pratique et intéressant ? Envoie-nous tes impressions, tes commentaires ou tes suggestions. Tu peux les adresser au bureau du Commissaire aux langues officielles, à Ottawa. Bonne année scolaire !

2

COMMENT GÉRER TON TEMPS

HOW TO MANAGE YOUR TIME

JUNE JUIN

Sunday Dimanche	Monday Lundi	Tuesday Mardi	Wednesday Mercredi	Thursday Jeudi	Friday Vendredi	Saturday Samedi
						1
2	3	4	5	6	7	8
9	10	11	12	13	14	15
16	17	18	19	20	21	22
23/30	24	25	26	27	28	29

MOST CANADIANS MAY LIVE IN CITIES, BUT EVERY CANADIAN KNOWS THE MAGIC OF OUR MOUNTAINS, THE HUSH OF OUR FORESTS AND THE POWER AND PEACE OF OUR RIVERS AND LAKES. PARKS CANADA HAS SET ASIDE REGIONS OF WILDERNESS FOR OUR USE, SAFE FROM THE INVASION OF INDUSTRY, BUSINESS OR "CIVI-LIZATION". YOU CAN FIND OUT ABOUT OUR NATIONAL PARKS SYSTEM BY CONTACTING PARKS CANADA. THEY'LL GIVE YOU ALL THE INFORMATION YOU NEED IN EITHER ENGLISH OR FRENCH.

BIEN QUE LA MAJORITÉ DES CANADIENS ET DES CANADIENNES SOIENT DES CITADINS, LA MAGIE DE NOS MONTAGNES, LE SILENCE DE NOS FORÊTS ET LA FORCE SEREINE DE LACS ET DE NOS RIVIÈRES NE LES LAISSENT PAS INDIFFÉRENTS. PARCS CANADA A CRÉÉ DES RÉSERVES NATURELLES POUR QU'ILS PUISSENT EN PROFITER, LOIN DE LA "CIVILISATION", DE SES USINES ET DE SES GRATTE-CIEL. TU PEUX OBTENIR DES RENSEIGNEMENTS SUR SON RÉSEAU DE PARCS NATIONAUX EN FRANÇAIS OU EN ANGLAIS.

---

⑥

PERSONAL COMMITMENT

RÉSOLUTION PERSONNELLE

Earlier in **1** and **2**, you may have identified an area in need of improvement as well as your own personal time wasters. Here is a chance to make a contract with yourself and improve a specific skill or add a new practice to your work habits. Remember to be S.M.A.R.T.

Tu en peux-être relevé en **1** et en **2**, un aspect à améliorer ainsi qu'une habitude qui te fait perdre du temps. Profite de cette occasion pour décider comment tu peux changer certains comportement et te faciliter la tâche. N'oublie pas les principes du travail SENSÉ.

"Do you think the teacher will notice I wrote my assignment on the bus?"

« Penses-tu que le professeur va s'apercevoir que j'ai écrit mon devoir dans l'autobus ? »

12

Mid-Term Exam
Examen semestriels

SUBJECT/MATIÈRE	DATE	LOCATION/LIEU

Examens finale
Final Exams

MATIÈRE/SUBJECT	DATE	LIEU/LOCATION

13

APOGEE DESIGNS – BROCHURE

DESIGN FIRM
GRAFFITO

ART DIRECTOR
TIM THOMPSON

DESIGNER
JOE PARISI

PHOTOGRAPHER
ED WHITMAN

WRITER
TONY MAFALE

CLIENT
APOGEE DESIGNS

Apogee Designs—a manufacturer of engineering materials— asked Graffito to design a promotional piece demonstrating its plastic-forming capabilities. After using pencil sketches to develop an approach for the piece, Graffito prepared a presentation comp on its computer. This included simulating the plastic-formed cover.

Apogee assisted in the production process through a variety of techniques, such as using synthetic paper, plastic spiral binding, and a clear matte plastic cover with vacu-formed graphic silhouettes.

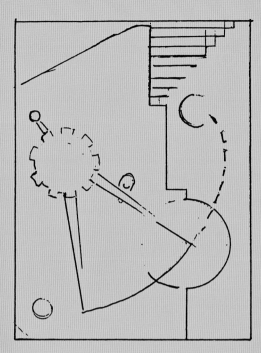

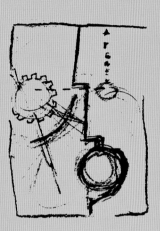

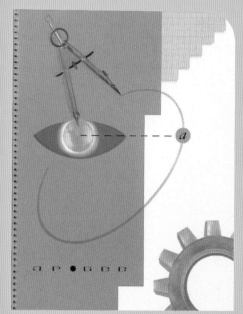

a p • g e e

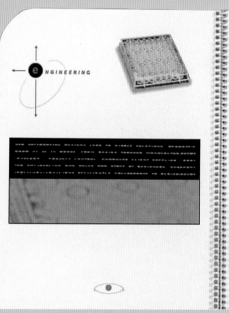

ENGINEERING

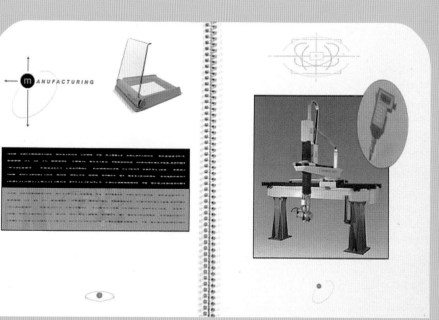

ANUFACTURING

ARLINGTON INTERNATIONAL RACECOURSE – BOOKLET

DESIGN FIRM
PLANET DESIGN COMPANY

ART DIRECTORS
KEVIN WADE, DANA LYTLE

DESIGNER
KEVIN WADE

ILLUSTRATOR
ARDEN VON HAEGER

CLIENT
ARLINGTON
INTERNATIONAL
RACECOURSE

Planet Design Company was given one month to design and an additional week to scan, print, and deliver an introductory booklet on thoroughbred racing for the Arlington International Racecourse. This schedule meant that no time existed for shooting transparencies, so delicate art had to be scanned directly. Planet Design was able to save some production time by selecting illustrations over photographs, although one illustration was rejected by the client after it was finished.

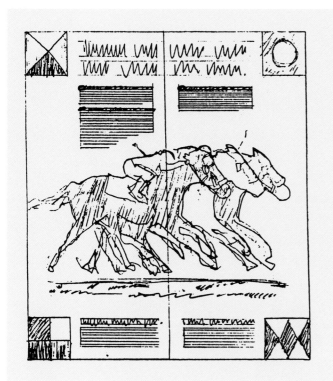

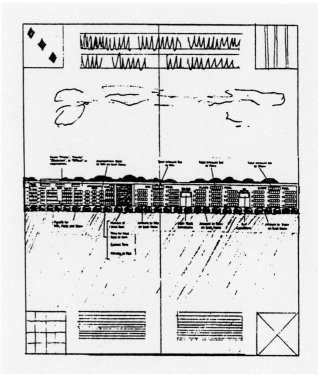

ART IN BLOOM

DESIGN FIRM
PLANET DESIGN COMPANY

**ART DIRECTORS /
DESIGNERS**
KEVIN WADE, DANA LYTLE

CLIENT
MADISON ART CENTER

The Madison Art Center asked Planet Design Company to create an awareness poster for its 1990 spring celebration, "Art in Bloom." The center wanted an image that could appear not only on the poster but on invitations, programs, and complimentary bookmarks as well.

The $500 budget was essentially enough to cover the studio's out-of-pocket expenses. Costly four-color scans and film work were not an option, so public domain images and photocopies were used. Vendors donated their services, and all the items were run together on a 26- by 40-inch press sheet to save on the printer's costs.

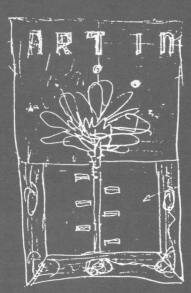

Art in

Asea Brown Boveri – Identity

ASEA BROWN BOVERI

DESIGN FIRM
PENTAGRAM DESIGN

DESIGNERS
ALAN FLETCHER,
QUENTIN NEWARK

CLIENT
ASEA BROWN BOVERI

Asea Brown Boveri, an international electrical engineering company, was formed when Asea, a Swedish firm, and Brown Boveri, of Switzerland, merged. It commissioned Pentagram to create a logo and corporate identity. The new company, however, was unsure whether to call itself AB, BA, ABB, etc., so Pentagram had to proceed on several fronts. Once ABB was selected, only two weeks were left before a completed design was due. Pentagram's Alan Fletcher and Quentin Newark divided each initial of the new name into quarters to represent the company's operation on four main business areas: power plants, power transmission, power distribution, and industrial equipment.

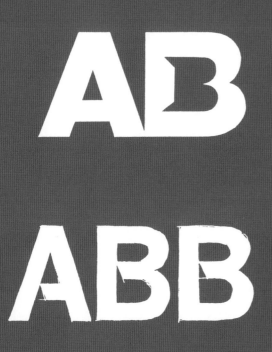

13

AVIATECA — IDENTITY

AVIATECA ⬤

DESIGN FIRM
BRIGHT & ASSOCIATES

ART DIRECTOR
KEITH BRIGHT

DESIGNERS
RAYMOND WOOD,
BARBARA EADIE

CLIENT
AVIATECA

Aviateca, Guatemala's national airline, asked Bright & Associates to examine the Central American travel market, then prepare a new graphic identity package including a logo and various applications. Aviateca wanted Bright's help at reversing the long-standing negative perceptions of it and at increasing receptivity in foreign markets, especially the United States.

After investigating perceptions of both Guatemala and Aviateca, the Bright study team recommended that the airline counter Central America's less positive associations with a graphic image from the area's rich cultural heritage.

AVIATECA

AVIATECA

AVIATECA

AVIATECA

Aviateca

Λviateca

Aviateca

AVIATECA

The Bright designers adapted a Central American pyramid form as the new Aviateca logo, then enhanced it with colors and patterns from Guatemalan crafts. Guidelines were prepared for applying the logo to aircraft, surface vehicles, and printed materials. The designers also oversaw the development of coordinating uniforms for Aviateca's in-flight and land staff.

The resulting identity appeals to the casual tourist by suggesting warmth and vitality, while implying the service and professionalism demanded by the business traveler.

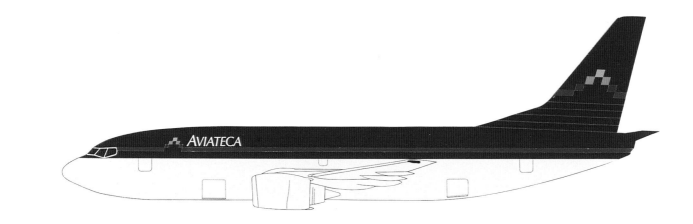

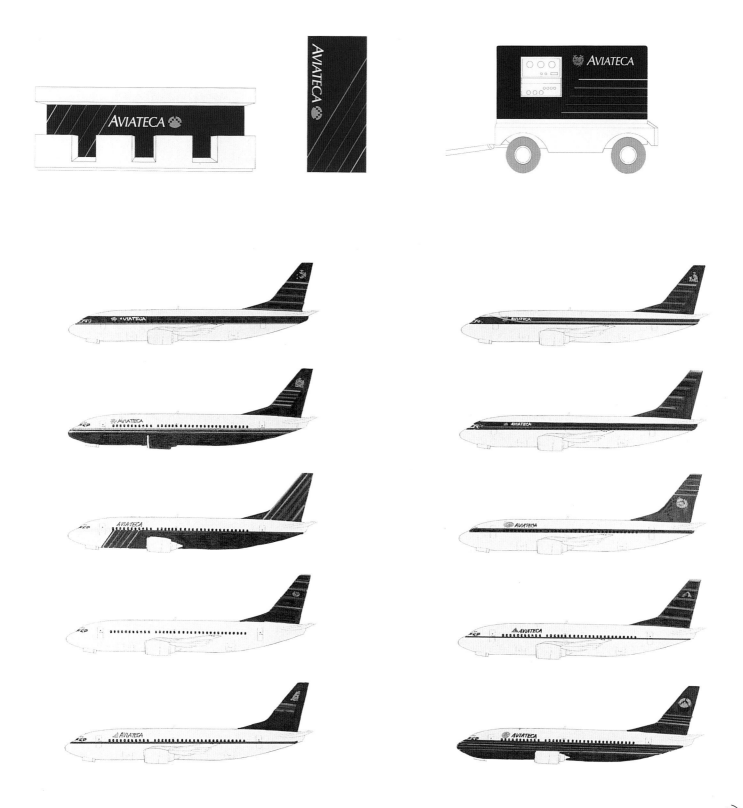

BISCOTTI NUCCI – PACKAGING

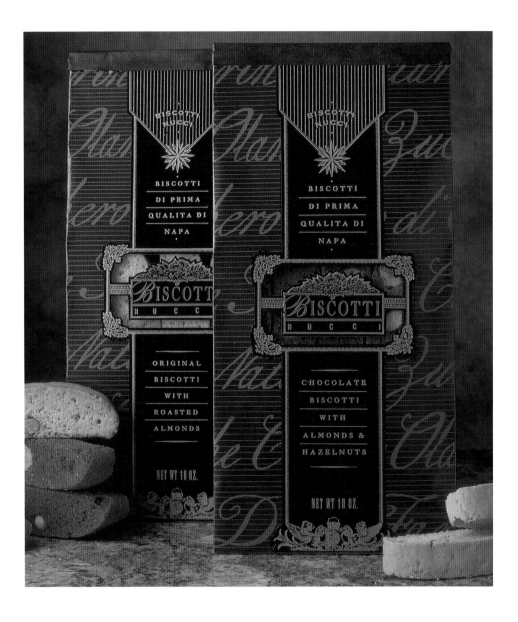

DESIGN FIRM
MORLA DESIGN

ART DIRECTOR
JENNIFER MORLA

DESIGNERS
JENNIFER MORLA,
JEANETTE ARAMBURU

ILLUSTRATOR
JEANETTE ARAMBURU

WRITER
ANNI MINUZZO

CLIENT
PRODOTTI NUCCI

Biscotti Nucci are original Venetian-recipe Italian cookies that are sold at high-end specialty department stores and food emporiums. Morla Design's research found that packaging for other biscotti brands utilized brightly colored type against a vast white field. Wishing to make Biscotti Nucci's product distinctive, attractive enough to purchase as a gift, and reflective of the recipe's heritage, Morla Design created a packaging system using rich metallic golds reversing out of a wide black band, accented by the engraved Venetian motifs of the lions and the sun.

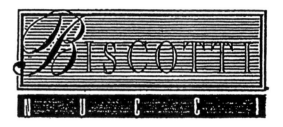

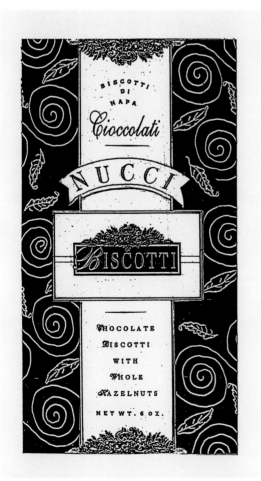

BLACKHAWK GRILLE – IDENTITY

DESIGN FIRM
THARP DID IT

ART DIRECTOR
RICK THARP

DESIGNERS
JANA HEER, JEAN
MOGANNAM, KIM
TOMLINSON, RICK THARP

INTERIOR DESIGN FIRM
ENGSTROM & HOFLING

FABRICATOR
ECLIPSE (RUSS WILLIAMS)

**PHOTOGRAPHER FOR
PRODUCT SHOTS**
KELLY O'CONNOR

CLIENT
BLACKHAWK GRILLE
(CALIFORNIA RESTAURANT
GROUP)

The Tharp Did It design studio was commissioned to develop a graphic identity for the Blackhawk Grille in San Francisco. Art director Rick Tharp and his associates combined automotive references and eclectic design elements in creating a look for the restaurant, which was near a classic car museum. A "BG" ligature with wings and tail feathers of a hawk was featured throughout the restaurant, including on the host station and the door handles.

Because of tight scheduling, Tharp discussed all preliminary design work with the interior designers. He then went directly to blueprints, at which point the client first saw the design.

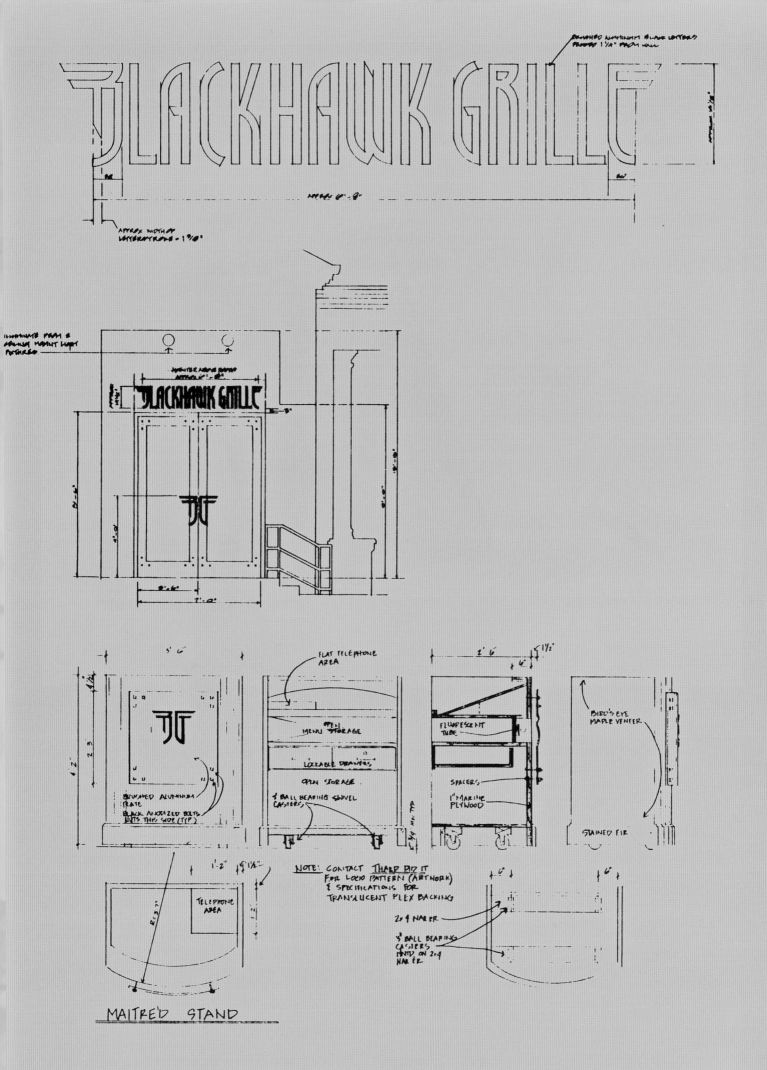

MAITRE'D STAND

NOTE: CONTACT THABR BIZ IT
FOR LOGO PATTERN (ARTWORK)
& SPECIFICATIONS FOR
TRANSLUCENT PLEX BACKING

On a visit to the client's office, after completion of the original hand-lettered logotype, Tharp found a copy of the logotype pinned to a corkboard with a handwritten note saying "Save the Grillf." He then realized that there was a problem with the legibility of the letter "E." Initial printings of the stationery and menus had been completed, but most would be reprinted within the next few months. The client did not want to change the logotype. Nevertheless, Tharp's designers redrew the "E" for all future applications, and the studio produced mechanicals for all printed materials at its own expense.

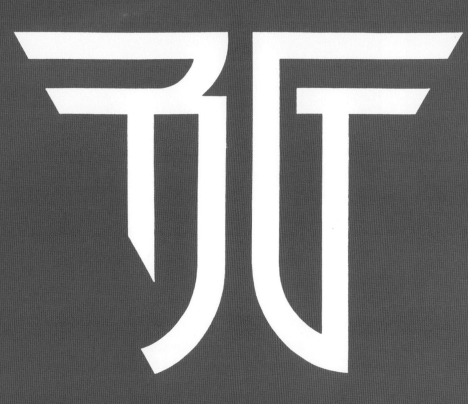

Tharp Did It fabricated a wine-by-the-glass board from a sheet of galvanized aluminum so that listed wines could be changed daily with a dry wipe marker. The logo and ruled lines were silk-screened onto the face, while the frame was built into the wall and covered with the same handmade paper as was used for the walls. The designers also developed coordinated labels for bottles of the restaurant's house wines.

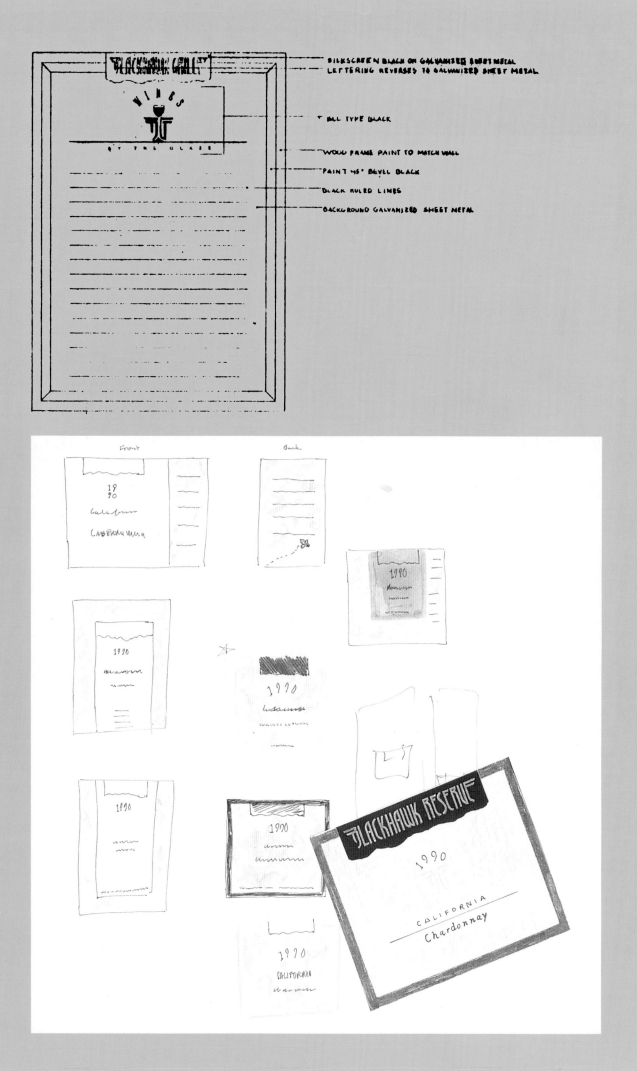

SILKSCREEN BLACK ON GALVANIZED SHEET METAL
LETTERING REVERSES TO GALVANIZED SHEET METAL

ALL TYPE BLACK

WOOD FRAME PAINT TO MATCH WALL

PAINT 45° BEVEL BLACK

BLACK RULED LINES

BACKGROUND GALVANIZED SHEET METAL

Front

Back

1990

CALIFORNIA

1990

1990

1990

1990

1990

1990

CALIFORNIA

BLACKHAWK RESERVE

1990

CALIFORNIA

Chardonnay

BOLD HOLD – PACKAGING

DESIGN FIRM
JOED DESIGN

CREATIVE DIRECTOR
EDWARD REBEK

DESIGNERS
JOANNE REBEK, ED REBEK

CLIENT
ALBERTO CULVER
COMPANY; JOVO BALACH,
PAUL NEMEJC, HAROLD
EFFINGER

Alberto Culver Company asked JOED Design to develop limited-edition packaging for its Bold Hold hair-care products. JOED presented Alberto Culver with 17 different concepts, all of which utilized funky graphics, wild patterns, and hot, neon colors. After teen focus groups chose three approaches as their favorites, Alberto Culver decided to produce all three designs for use with different Bold Hold products.

After the new packaging was introduced, Bold Hold's sales increased dramatically. The result: Albert Culver permanently replaced its old packaging with the new designs.

FLIP-OVER
VOLUMIZING SPRAY

WORLD
91
ROCK

CIPANGO BEER – IDENTITY

DESIGN FIRM
BRIGHT & ASSOCIATES

ART DIRECTOR
KEITH BRIGHT

DESIGNERS
RAYMOND WOOD,
MARK VERLANDER

CLIENT
KIRIN BREWERY COMPANY

Kirin, a Japanese brewer, needed a name and identity for a beer it planned to market in California. It asked Bright & Associates to help position the beer for that state's highly competitive imported-beer market. In consultation with Hakuhodo, Kirin's advertising agency, Cipango—the name given to Japan by Marco Polo—was selected as the beer's name. This gave the product a European feel while retaining its Japanese heritage. To convey adventure, Kirin provided label copy alluding to the name's origin, and the Bright designers prepared a label featuring an old-style map and explorer's compass.

CLARIS CORPORATION – PACKAGING

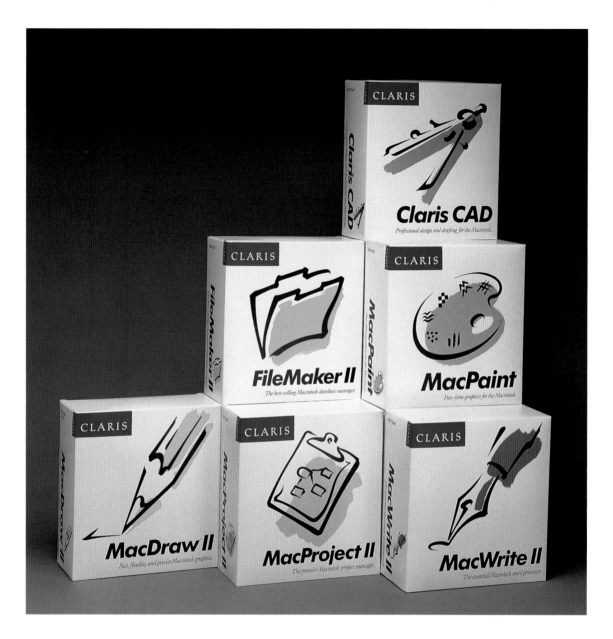

DESIGN FIRM
NEUMEIER DESIGN TEAM

DESIGNERS
MARTY NEUMEIER,
CHRISTOPHER CHU

ILLUSTRATOR
CURTIS WONG

WRITERS
DESIREE LAGRONE, MARTY
NEUMEIER

CLIENT
CLARIS CORPORATION

Claris Corporation asked the Neumeier Design Team to revamp its packaging to give its various products a bold look that would retain a uniform identity while bringing out each product's separate "personality."

Drawing from a combination of store research and corporate strategy, Neumeier Design proposed, and Claris agreed, to follow an approach that favored "simplicity over complexity, playfulness over seriousness, and accessibility over density." The studio's designers and illustrators brainstormed ideas for icons that would convey the central strength of each software project.

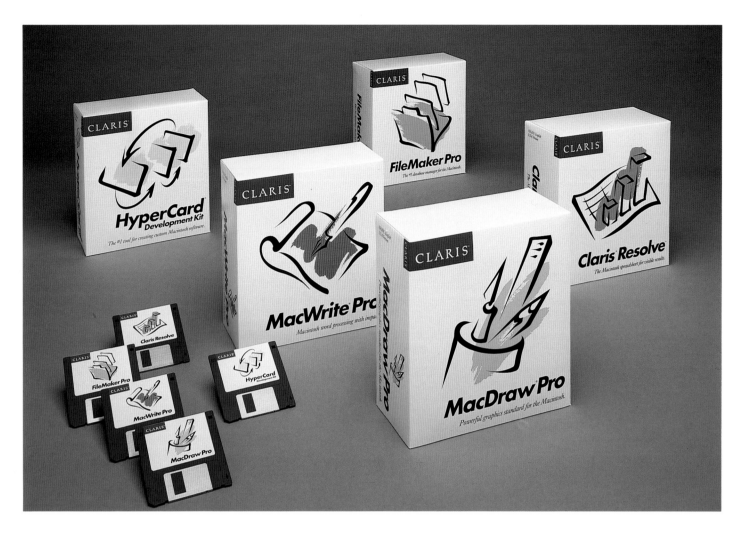

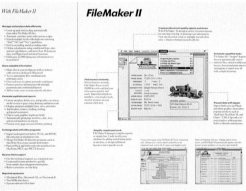

Some icons required several attempts as the designers and Claris searched for illustrations that were straightforward without being too simple or deceptive.

Neumerier Design had less than three months to develop and produce this look for its initial use with six software products. It has since been used for three more products introduced after the initial redesign was unveiled.

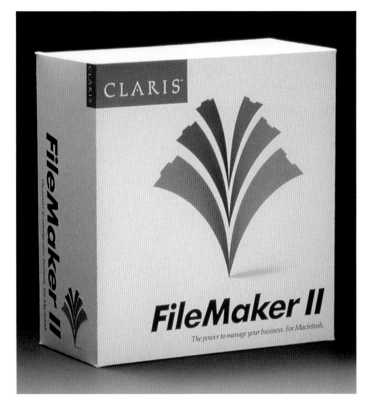

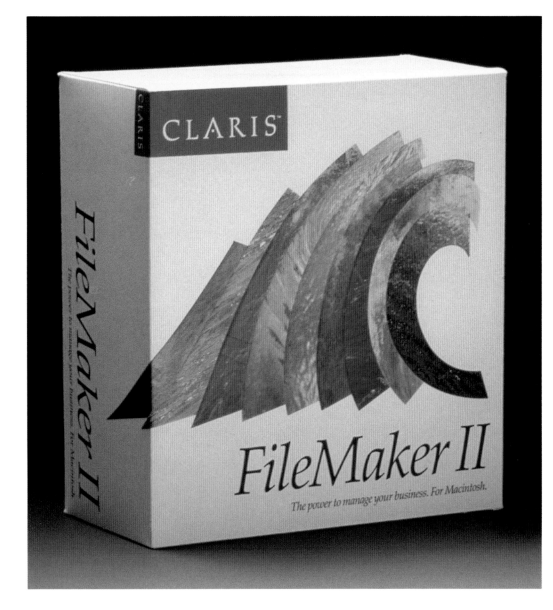

Concert Artists of Baltimore – Poster

Design Firm
Graffito

Art Directors
Tim Thompson,
Dave Plunkert

Designer
Dave Plunkert

Photographer
Stephen John Phillips

Writer
Concert Artists of
Baltimore

Client
Concert Artists of
Baltimore

Graffito began producing promotional materials for the Concert Artists of Baltimore's annual concert / auction by deciding to break with previous years' practice and prepare a poster for handing out at the event and a separate invitation. Perhaps because the studio was working pro bono, its proposal met little resistance from Concert Artists. Graffito recruited an appropriate photographer, prepared props for the photo, and designed the poster's graphics and typography around the photo.

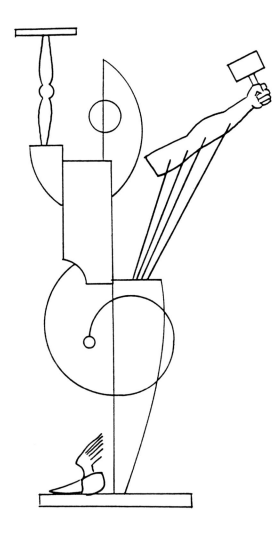

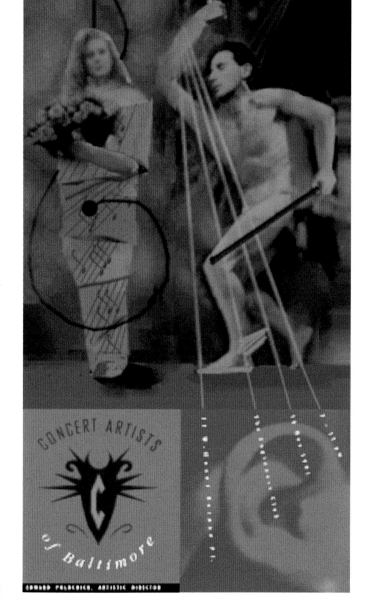

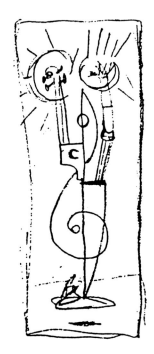

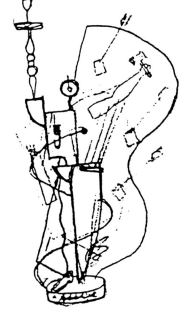

CUSTOM PRINTERS – STATIONERY

DESIGN FIRM
246 FIFTH DESIGN
ASSOCIATES

ART DIRECTOR
TERRY LAURENZIO

DESIGNER
TERRY LAURENZIO,
GREG TUTTY

ILLUSTRATORS
TERRY LAURENZIO,
JOANNE JEFFS

CLIENT
CUSTOM PRINTERS OF
RENFREW

Custom Printers of Renfrew asked 246 Fifth Design Associates to develop stationery that would appeal to creative agencies and studios in nearby Ottawa. After meeeting with printshop officials, 246 Fifth owner Terry Laurenzio prepared thumbnail sketches for several possible concepts. Once agreement was reached on an approach placing the "old" in a contemporary context, 246 Fifth prepared line illustrations that appeared on the stationery as "ghosted back images of the past." The design studio prepared camera-ready art, then Custom Printers handled the film work and printing.

DESIGNERS CHILI COOKOFF – POSTER

DESIGN FIRM
SIBLEY/PETEET DESIGN

ART DIRECTOR
DON SIBLEY

WRITER
DON SIBLEY

ILLUSTRATOR
DON SIBLEY

CLIENT
DALLAS DESIGN FIRMS

Each fall, Dallas design firms compete in the Designers Chili Cookoff. A poster advertising the competition is designed by the previous year's winner. Sibley/Peteet Design won the 1989 competition and therefore was responsible for the 1990 poster. The studio decided on an image of a chili pot cooking over an open fire. Smoke billowing from the pot was done as a loose, bold brush stroke splattered with bright colored paint. Instead of a stirring spoon, a brush was placed in the foreground. The art style overall was loose, gritty, and a little wild—capturing the personality of the event perfectly.

DRUKKERIJ MART. SPRUIJT BV — CALENDAR

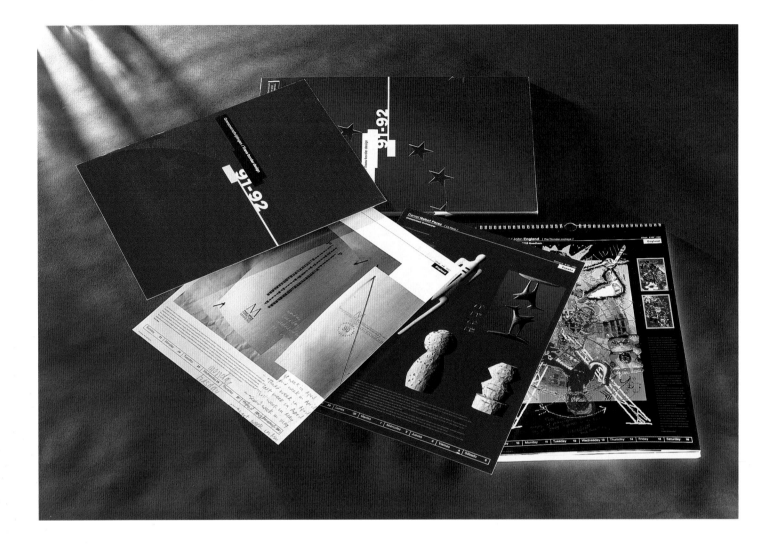

DESIGN FIRM
SAMENWERKENDE
ONTWERPERS

CONCEPT CREATORS
MARIANNE VOS, ANDRÉ
TOET

ART DIRECTOR / DESIGNER
MARIANNE VOS

CLIENT
DRUKKERIJ MART. SPRUIJT
BV

Netherlands-based design firm Samenwerkende Ontwerpers prepared the 1991 promotional calendar for the Dutch printer Drukkerij Mart. Spruijt bv. The calendar focused on European graphic, architectural, and interior designers who "flaunt the boundaries of their profession."

Once Samenwerkende Ontwerpers decided which designers to include, the firm recruited prominent Dutch writers to comment about their work. The calendar's typography was kept plain and structured, so as not to interfere with the featured design works.

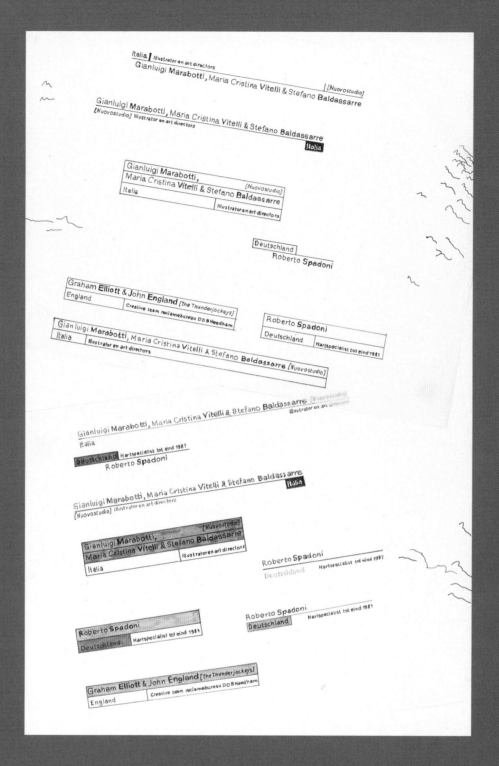

oktober 1991 week 19
maandag **6** dinsdag **7** woendag **8** donderdag **9** vrijdag **10** zaterdag **11** zondag **12**

Graham **Elliot** & John **England** THE THUNDERJOCKEYS

ENGLAND **Creative team reclamebureau DDB Needham**

Daniel Nebot **Perez**

ESPANA LA NAVE
Disenador

41

Esse – Portfolio

DESIGN FIRM
GRAFIK COMMUNICATIONS

CREATIVE DIRECTOR
JUDY F. KIRPICH

DESIGNERS
MELANIE BASS,
GREGG GLAVIANO

PHOTOGRAPHER
CLAUDIO VAZQUEZ

AIR SCULPTOR
BETSY THURLOW SHIELDS

CLIENT
GILBERT PAPER

Gilbert Paper commissioned Grafik Communications to prepare a presentation portfolio for its new Esse high-end papers. To highlight the papers as recycled and "natural," Grafik proposed a box made of recycled cardboard, secured with a stick and rope. Various sketches for box and lid designs were presented to Gilbert for selection.

Inside the box was placed an accordion-folded specifications card and samples of printed materials. The "natural" theme was continued throughout the portfolio with illustrations of sculptures composed of natural elements such as leaves, sticks, and seeds.

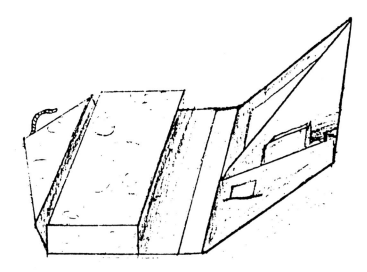
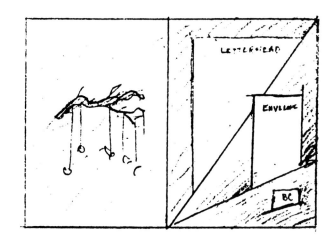

EventMedia International — Identity

DESIGN FIRM
MICHAEL DORET

ART DIRECTOR
MICHAEL DORET

CLIENT
EVENTMEDIA
INTERNATIONAL

EventMedia International sought a new graphic identity to suggest the breadth and quality of its entertainment and cultural clients. Designer Michael Doret envisioned classical graphic elements to suggest artistic arenas: a lyre for music, mask for drama, palette for art, and column for antiquities. With pencil sketches, he developed the general look he wanted and finessed it with ink variations. He then finalized the logo by preparing a precise pencil draft and versions of the logo in black-and-white and in five colors. To conclude the project, a complete stationery set was designed and printed.

Expo-Vienna — Identity

VIENNA & BUDAPEST ®

Design Firm
Pentagram Design

Designers
Alan Fletcher,
Thomas Manss

Client
Expo-Vienna AG

Expo '95, the 1995 World's Fair, will be held in Vienna, Austria, and Budapest, Hungary—the first such fair to be hosted by two cities and in two countries simultaneously. The fair's organizers, Expo-Vienna AG, sponsored an international competition to select a trademark and marketing symbol to convey the expo's theme of "Bridges to the Future." In December 1990, an international jury of graphic designers selected a symbol submitted by Pentagram Design Limited, a London-based firm with studios in the United Kingdom and the United States.

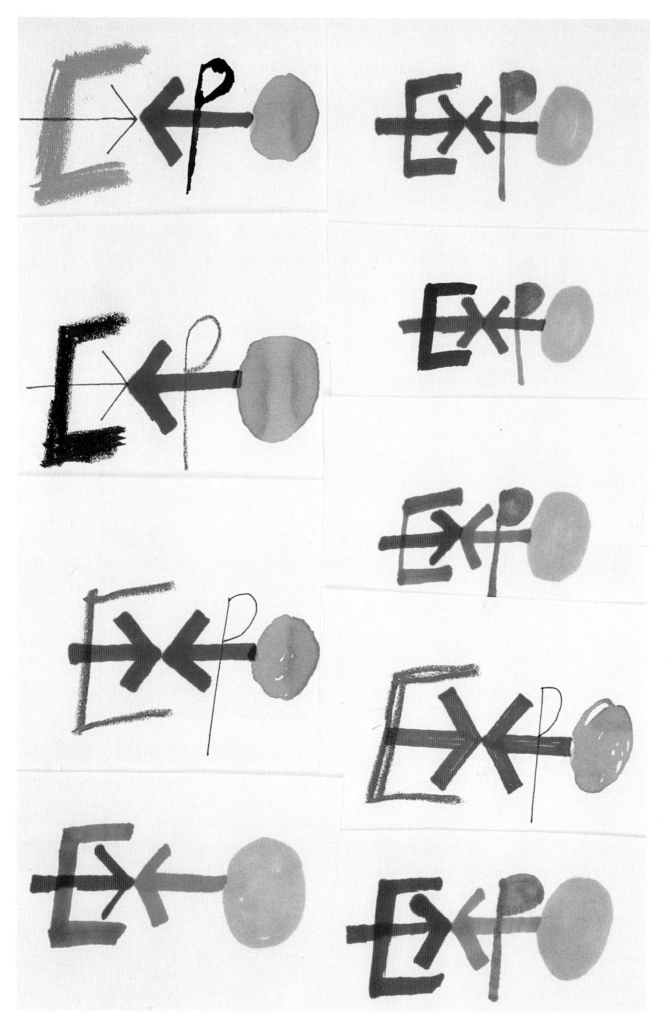

EYE OF THE SWAN — IDENTITY

DESIGN FIRM
THARP DID IT

ART DIRECTOR
RICK THARP

DESIGNERS
JANA HEER, KAREN
NOMURA, FRANK
POLLIFRONE, RICK THARP

ILLUSTRATOR
MICHAEL BULL

CALLIGRAPHER
GEORGIA DEAVER

**PHOTOGRAPHER FOR
PRODUCT SHOTS**
KELLY O'CONNOR

CLIENT
SEBASTIANI VINEYARDS

The original labeling for Sebastiani's Eye of the Swan did not reflect its delicate color or character. The Tharp Did It design studio was asked to develop new labeling for this blush wine and a companion product, Black Beauty. The studio created handlettered logotypes to accompany an illustration of a swan. On the back labels, Rick Tharp made the univeral pricing code (UPC) symbols an integral part of the design by graphically incorporating it into a field of cattails.

Since the firm's first label redesign in 1985, Tharp Did It has been called in every two years to update the label as part of Sebastiani's marketing efforts to keep the label "fresh and current."

EYE OF THE SWAN

FASHION ISLAND – IDENTITY

DESIGN FIRM
SIBLEY/PETEET DESIGN

ART DIRECTOR / DESIGNER
DON SIBLEY

ILLUSTRATORS
DON SIBLEY, TOM HOUGH

WRITER
DON SIBLEY

CLIENTS
DONAHUE/SCHRIBER,
FASHION ISLAND

Fashion Island—an upscale, mostly open-air shopping center in Newport Beach, California—selected Sibley/Peteet Design to create promotional materials for a major repositioning campaign. Art director Don Sibley centered his overall design scheme on creating a distinct graphic signature for the retail center. He developed an illustrative style vaguely reminiscent of Gauguin to portray the center as a Fantasy Island destination. The style was characterized by a brushstroke-like black line, actually executed with a marker, and loosely registered colored shapes to help define the elements.

Lorem ipsum dolor sit enim ad minimum veni dolor in reprehender dignissim qui blandit p culpa qui officia deser nobis eligend optio com repelend. Temporem Itaque earund rerum hic ne ad eam non possi ad augendas cum consc gen epular reliquqrd videantur, Invitat igitur

FARMERS MARKET

Husmo tempor incidunt ut labore et dolore magna aliquam laboris nisi ut aliquip ex ea commodo consequat. Duis illum dolore eu fugiat nulla pariatur. At vero eos et accusam molestais excepteur sint occaecat cupidat non provident, harumd dereud facilis est er expedit distinct. Nam liber aque placeat acer possim omnis es voluptas assumenda necessit atib saepe eveniet ut er repudiand sint

Dusmod tempor incidun laboris nisi ut aliquip illum dolore eu fugiat excepteur sint occaecat eiusmod tempor in

AMBIENCE

Amet, consectes amu quis nostr in voluptate rasesent luontum amet, consect illum dolore eu excepteur sint dolor in repre

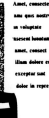

Dur ac ipscing elit, sed diam nu amy eiusmod tempor inc molestaie son consequat, vel illum dolore eu fugiat md dereud facilis est er expedit distinct. Nam liber a temp diand sint et molesua non aste recusand. Itaque earu ne ad eam non possi.ig accommodare nost ros quos tu cupidat, quas nulla praid im umdnat. Improb pary nag veling en liberalitat magnis em conveniunt. babut tu

Dunimim veniami quis nostrud exerci a prasesent lupatum delenit aigue duos do anim id quo maxid placit at lacer possim eiusmod tempor incidunt ut labore e laboris nisi ut aliquip ex ea commodo c dignissim qui blandit prasesent Itaque earund rerum hic teneturv

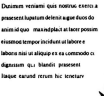

Atnaipscing elit, sed diam nonu d exercitation ullamcorpor suscipt t esse molestaie son consequat delenit aigue duos dolor et me

DINING

FASHION

Map to The Island

GARDEN GROVE FWY
SAN DIEGO FWY
NEWPORT CENTER
FASHION ISLAND

Ad diam nonumy eiusmod tempor incidunt ut vel illum dolore eu fugiat nulla pariatur. t er expedit distinct. Nam liber a tempor cum s a non aste recusand. Itaque earund rerum hic tene augendas cum cr nscient to lactor tum toen lea vera ratio bene san santos ad iustitiami aequitatd

FASHION ISLAND
NEWPORT CENTER

ENTERTAINMENT

FASHION ISLAND
NEWPORT CENTER

orem ipsum dolor sit amet, con

olor in repre mider it in voluptate velit esse moles

culpe qui officia deserunt mollit anim id est l

repellend. Temporem autem quinsua et sur office

ne ad eam non possing accommodare nost ros

gen epular reliquqrd on cupidat, quas nulla praid im

illum dolore eu fugiat nulla pariatur. At vero eos

harumd dereud facilis est er expedit distinct. Nam li

necessit atib saepe eveniet ut er repudiand sint eLm

memorite it tum etia ergat. Nos amice et nebevol ol

minuti potius inflammad ut coercend matis

molestais excepteur sint occaecat cupiosi non provide.

The resulting overall look was bright, fresh, and unstudied, projecting a sense of sophistication and fashion flair—representative of the center. The shopping center used the campaign for two years, incorporating it into all sorts of advertising and printed support materials.

FRANK PARSONS PAPER – PROMOTIONAL KIT

DESIGN FIRM
SUPON DESIGN GROUP

ART DIRECTOR
SUPON PHORNIRUNLIT

DESIGNER
ANDREW DOLAN

PHOTOGRAPHER
EARL ZUBKOFF

CLIENT
FRANK PARSONS PAPER
COMPANY

Frank Parsons Paper Company asked Supon Design Group to prepare a kit promoting the wholesaler's coated papers to printers. Parsons suggested using a picture of its building, but art director Supon Phornirunlit persuaded them that a look conveying "coated" would be more appropriate. His designers explored images that were in some way coated—candies, paint, a tuxedo jacket, etc.—then decided that the kit's audience might best relate to doughnuts. Inside the kits were printed sheets of various paper stocks, with each sheet showing how portions of the same photo reproduced with and without gloss varnish; and using four-color, duotone, and halftone processes.

FLAPS TUCK INTO DIE-CUTS
UNDER BOW TIE

INDIV.
PAPER
SAMPLES
CAN BE
SHIRTS

Futura Typeface – Poster

Design Firm
Morla Design

Designer / Illustrator
Jennifer Morla

Client
Mercury Typography

A San Francisco Bay-area typesetter asked designer Jennifer Morla to prepare a poster illustrating her favorite typeface. Selecting Futura—developed in 1930 by Paul Renner of the Bauhaus school as the first sans-serif typeface—Morla designed this poster to create an appreciation of the face, its history, and architectural use. The quotation is from the era's DeStijl artistic movement, which embraced Futura as "the typeface of the future." Morla featured Manhattan's Chrysler Building as the poster's graphic focus because of its use of Futura for signage. Morla used a vertical format for the poster to reinforce the architectural sense of height.

Applications of
Futura at the time?

The object of nature
is man
The object of man
is style
(De Stijl quote)

Copy pertaining
to How Moderne,
De Stijl and
Bauhaus all
embraced Futura
as the typeface of the
"future"

Investigate
pos/neg relationships

date?
verticle

negative?

positive?

De Stijl quote

Architectural
application of
typeface

GSD&M – Identity

Design Firm
Sibley/Peteet Design

Creative Director
Tim McClure (GSD&M
agency principal)

Art Director
Rex Peteet

Designers
Rex Peteet, Julia
Albanesi

Client
GSD&M

ibley/Peteet was commissioned to design a new graphic identity for GSD&M. The objective was a look that would represent the varying personalities of the advertising agency's four original principals, whose last initials formed the firm's name. Sibley/Peteet redesigned four classic typefaces— Garamond (borrowed and modified from GSD&M's existing logotype), Futura, Lubalin Graph, and Melior.

The ampersand in the agency's name was eventually omitted from the logo, because the designers felt it was implied even when not present.

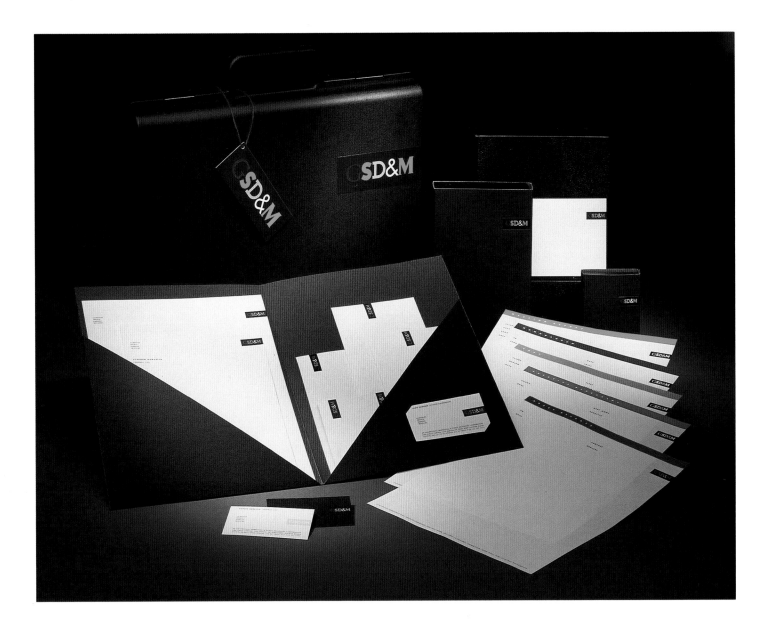

Classic black and hot red were designated the dominant colors in the palette—to signify, respectively, GSD&M's serious business character and its creative intensity. The other primary colors were used minimally (primarily with letters of the logotype) for accents and color coding.

The new graphic identity has been applied to various materials GSD&M produces to promote itself—from stationery and folders to attaché cases and the corporate flag.

GRAMP'S WINES — IDENTITY

DESIGN FIRM
BARRIE TUCKER DESIGN
PTY

DESIGNER
BARRIE TUCKER

TYPOGRAPHERS
BARRIE TUCKER,
ELIZABETH SCHLOOZ

PRODUCTION MANAGER
SERGIO JELOSCEK

CLIENT
G. GRAMP & SONS

When G. Gramp & Sons—an Australian vintner since 1847—decided to launch a new range of wines under the "Gramp's" label, it wanted a design reflecting the Gramp heritage while also present-

ing upmarket elegance. The company agreed to designer Barrie Tucker's proposal for a typographical design highlighting the Gramp's name. Tucker prepared a two-label approach. A single logo label was prepared for

all wines in the Gramp's line. Positioned below the Gramp's name was a descriptive label for each type of wine, with the name of each wine assigned its own ink color.

GRAMPS TYPE EMBOSSED & Gloss Varnish

Matt Finish Background etc.

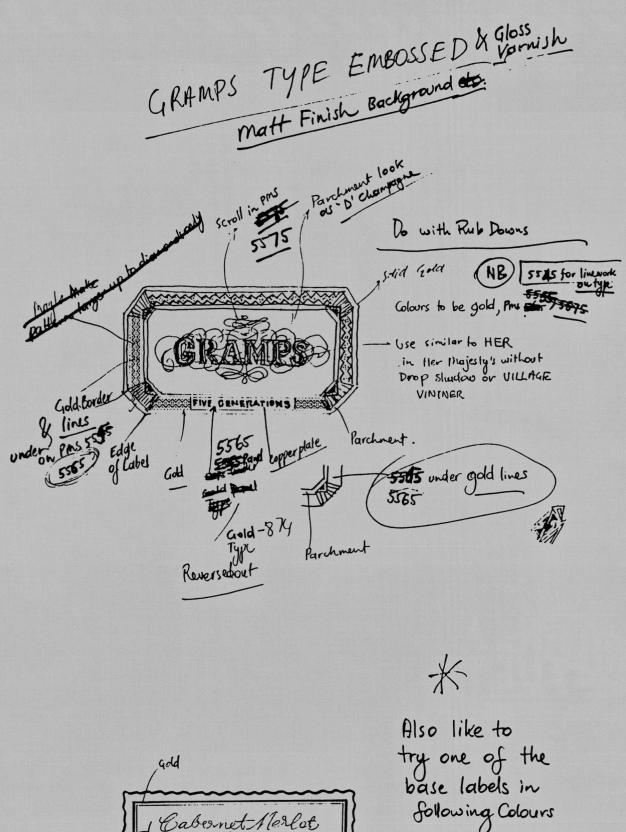

Scroll in Pms 5575

Parchment look as "D' Champagne".

Do with Rub Downs

Solid Gold

(NB) 5545 for linework on type.

Colours to be gold, Pms. 5575

Use similar to HER in Her Majesty's without Drop shadow or VILLAGE VINTNER

Gold Border & lines under on PMS 5585 5585

Edge of Label

Gold

5565 and copperplate

Gold Type

Gold-874 Type Reversed out

Parchment.

5585 under gold lines 5565

Parchment

Gold

229

Cabernet Merlot
Vintage 1987
WINE PRODUCT OF AUSTRALIA
· 750ml ·

5575

Also like to try one of the base labels in following Colours

229 Type Cabernet Merlot Vintage

on Parchment Background flat Paper

Borders (Gold) - Product of Aust & 750ml. 5575

PLEASE

HILLSIDE NEIGHBORHOOD – IDENTITY

DESIGN FIRM
SAYLES GRAPHIC DESIGN

ART DIRECTOR
JOHN SAYLES

CLIENT
HILLSIDE DEVELOPMENT
CORPORATION

Designer John Sayles was asked to develop an identity package for the proposed Hillside Neighborhood. A visit by Sayles to the future site of the development gave him an idea of a place where "magnificent gates open up to let you in."

When Sayles drafted a logo suggesting estate gates, the client requested two small changes: make the trees less whimsical and "Hillside" more legible. Sayles accomplished this by eliminating the white space within the "S" and disconnecting the

top of the second "I" from the gate-like structure. He also designed accompanying materials, including a desk pen set and coordinated packaging.

HILLSIDE

NEIGHBORHOOD

JAGUAR CAT-A-LOG – PROMOTIONAL BROCHURE

DESIGN FIRM
SIBLEY/PETEET DESIGN

ART DIRECTOR
DON SIBLEY

ILLUSTRATORS
DON SIBLEY, KELLY
STRIBLING SUTHERLAND,
DAVID BECK, JOHN EVANS,
DAN PICASSO, GARY
BASEMAN, JIM JACOBS,
TOM HOUGH, SEYMOUR
CHWAST, JACK UNRUH,
MICK WIGGINS, STUART
ASHMAN

PHOTOGRAPHERS
GILLES LARRAIN, RICK
DUBLIN, GREGORY
HEISLER, CHET
MORRISON, JOHN
PINDERHUGHES, RICHARD
KLEIN, MIKE WILSON

WRITER
JOHN FRAZIER

CLIENT
WEYERHAEUSER PAPER
COMPANY

The Jaguar Cat-A-Log is a follow-up promotional brochure featuring Weyerhaeuser Paper's Jaguar paper line. Weyerhaeuser commissioned Sibley/Peteet Design to develop a piece that would reinforce the attributes of this premium white sheet to the designer/specifier.

Art director Don Sibley's solution was the "Cat-A-Log." He designed a cover with no type and a double gatefold, which allowed for four very different cat portrait interpretations. The promotion was mailed in an envelope with the word *Jaguar* die-cut from its face. This revealed hints of color from the brochure inside.

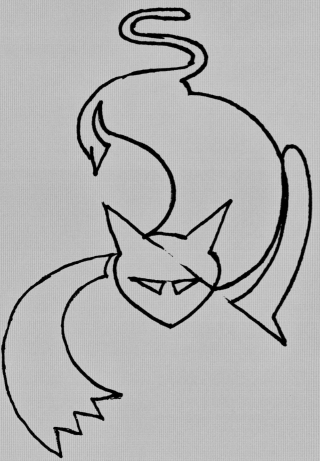

The "Cat-A-Log" was intended to poke fun at some common cat colloquialisms and highlight one of the most famous felines of all—the jaguar. To demonstrate the paper's capabilities, Sibley recruited an eclectic mix of illustrators and photo-graphers, then utilized a myriad of printing and finishing techniques.

All in all, the tone was meant to be playful and to entertain the designer audience while subtly rein-forcing name recognition for Weyerhaeuser's top cat.

JUCHHEIM'S IMAGE D'OR — IDENTITY

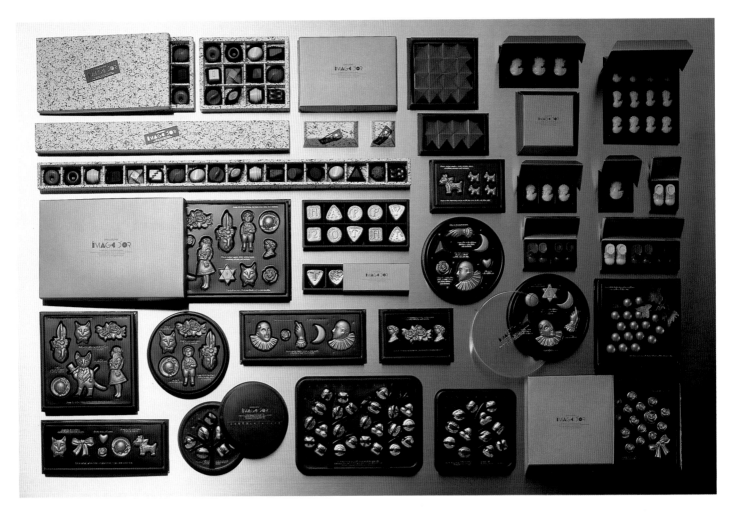

chocolatier
ÍMAGE D'OR

DESIGN FIRM
PAOS (MOTOO NAKANISHI,
CHAIRMAN; YUTAKA SANO,
EXECUTIVE DIRECTOR,
DESIGN DIVISION)

CLIENT
JUCHHEIM'S CO.

Juchheim's, a Japanese confectioner, commissioned PAOS to help market its new line of chocolates. To convince Japanese consumers that adults as well as children could enjoy chocolate, PAOS developed a strategy presenting the chocolates as a "personal gift" to be given by connoisseurs. It recommended "Image d'Or" ("Golden Image") as the brand name and designed chic packaging and an elaborate gift-wrapping service to help advance the product's image. PAOS also helped Juchheim's explore nontraditional shapes for the chocolates, such as architectural pieces, fairy tale characters, and delicate cameos.

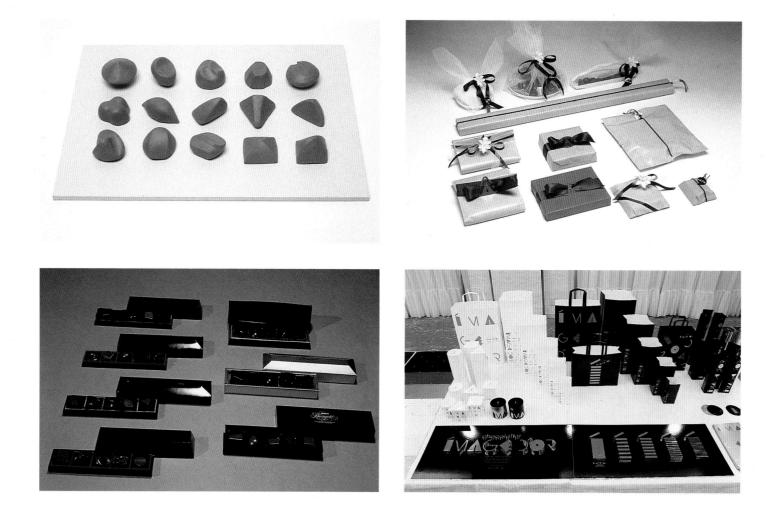

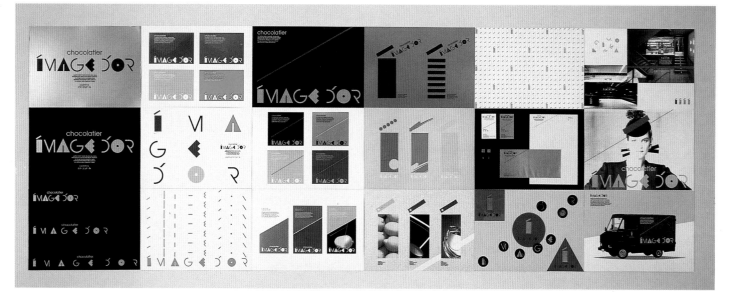

ki Research – Logo

ki research

Design Firm
Supon Design Group

Art Director
Supon Phornirunlit

Designer
Andrew Dolan

Client
ki Research

Ki Research, a computer software company, commissioned Supon Design Group to develop a company logo that would be modern, high-tech, and upbeat. The Supon designers rejected an initial approach drawing on *ki*'s meaning "tree" in Japanese, deciding that it did not convey an appropriate image for the company. Focusing instead on the company name, the designers showed the client five sketches using different renditions of "ki." The client asked them to further explore a sketch in which the top arm of the "k" pointed to the dot of the "i"—asking that the "k" be made more lively. When the designers returned with four new sketches, ki Research selected the option that conveyed the most movement.

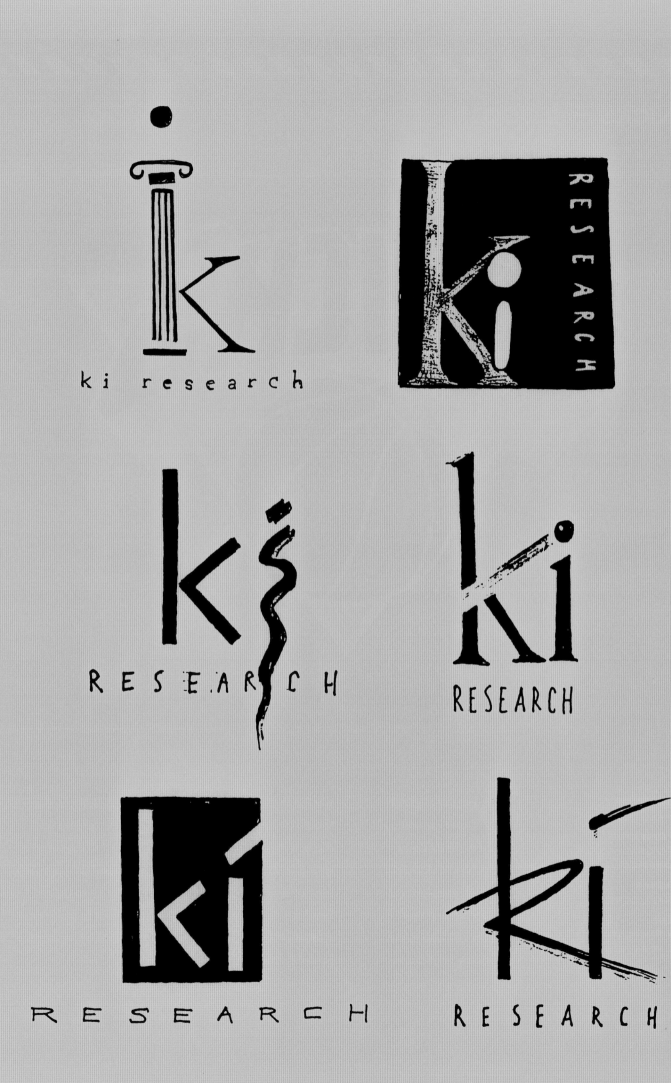

Knicks – Logo

TM

DESIGN FIRM
MICHAEL DORET

ART DIRECTOR
TOM O'GRADY / NBA
PROPERTIES

DESIGNER
MICHAEL DORET

CLIENT
NBA PROPERTIES /
NEW YORK KNICKS

Designer Michael Doret was commissioned to develop a new logo for the New York Knicks basketball team—one that would lend itself more easily than the old logo to merchandising applications. It was felt that black and silver should be incorporated into the new logo. There was also strong feeling at the beginning of the project that the new logo should include a symbol of New York City, probably the Empire State Building. In the end, however, this idea was discarded to simplify the design, and the approach most similar to the old logo was chosen to maintain continuity.

ABCDEFGHIJKLM
NOPQRSTUVWXYZ
1234567890 & A E I

Design Firm
Tharp Did It

Art Director
Rick Tharp

Designers and Illustrators
Jana Heer, Susan Jaekel, Jean Mogannam, Rick Tharp, Linda Woodruff

Photographer for Product Shots
Kelly O'Connor

Client
Le Boulanger

Le Boulanger hired the Tharp Did It design studio to update its logo and packaging to accompany a new look being prepared for the San Francisco Bay-area bakeries. The Tharp associates created a hand-drawn logotype that retained visual similarity to the existing logotype. They also designed an emblem around the logotype for use on packaging and labels. Many sketches were required before a baker character for the emblem was approved by the client. In addition, the designers created a custom stencil-styled typeface for use on the emblem as well as on other print materials and interior and exterior signage.

LeBoulanger™
—THE BAKER—

In designing new packaging, Tharp suggested that a separate bag with its own character be designed for each type of bread that Le Boulanger sold. After seeing rough sketches of Tharp's ideas for the bags, the client trusted the designers to proceed without further discussion or comps. Each bag featured a photograph of the type of bread inside and also incorporated the logotype or emblem, as well as a seal identifying Le Boulanger as an award-winning sourdough bakery. The logotype and emblem were used in different ways on each bag, but they all retained a related look.

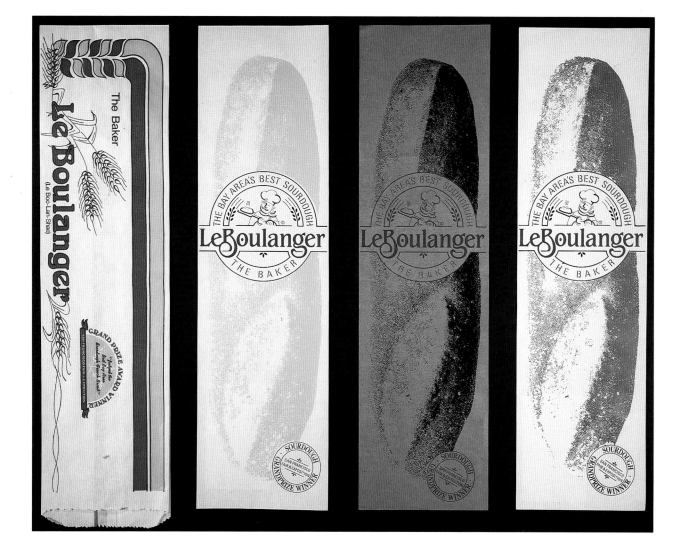

FRENCH ROUND

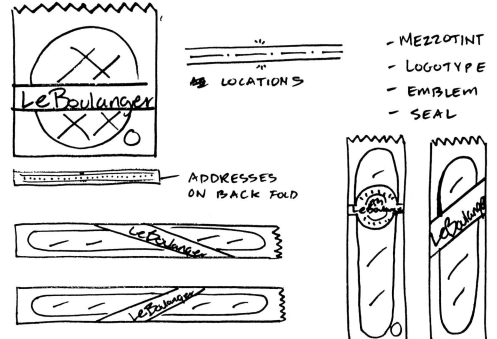

LOCATIONS

ADDRESSES
ON BACK FOLD

BAGUETTE

- MEZZOTINT
- LOGOTYPE
- EMBLEM
- SEAL

FRENCH LOAF

LEVI'S SILVER TAB – PROMOTIONAL BROCHURE

DESIGN FIRM
MORLA DESIGN

ART DIRECTOR
JENNIFER MORLA

PHOTOGRAPHER
DAVID MARTINEZ

STYLIST
JULIE DODGE

WRITER
JULIE KLEE

CLIENT
LEVI STRAUSS AND CO.

Designer Jennifer Morla had about five weeks to produce all aspects of this trade brochure for Levi's Silver Tab Collection—including design concept, photography, preparation, and printing.

Lifestyle photography in silver and black inks was used to create an elegant look, one that would appeal to wholesale buyers at department and clothing stores. The brochure was also meant to be informative, detailing the different styles and garment features of the Silver Tab Collection. Each photo in the brochure was also available for purchase by merchants as a foam-mounted poster ready for storeroom display.

— STUDY ONE

— STUDY TWO

Levi's® Silver Tab™ Collection

— STUDY THREE

MacArthur Park – Menus

DESIGN FIRM
MORLA DESIGN

ART DIRECTOR
JENNIFER MORLA

DESIGNERS
JENNIFER MORLA,
JEANETTE ARAMBURU

WRITER
JUDY RADICE

CLIENT
SPECTRUM FOODS

The MacArthur Park restaurants in San Francisco and Palo Alto, California, are popular places for people to gather and meet. The interior is constructed of exposed brick with a changing collection of fine art. The tables are covered with butcher paper with crayons available for customers to express their creativity.

When Jennifer Morla was asked to design the restaurants' menus, she wanted to enhance the establishments' interiors and reflect their ambiance. The result was the loose illustration of people gathering, which she combined with distinctive type and borders to create an image of combined casualness and elegance.

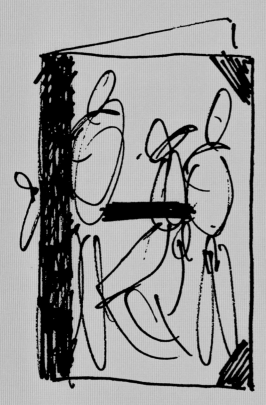

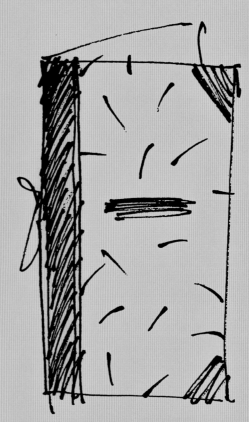

MacArthur Park

MacArthur Park

MacArthur Park

MACARTHUR PARK

MADISON AIDS SUPPORT NETWORK – POSTER

DESIGN FIRM
PLANET DESIGN COMPANY

ART DIRECTORS
KEVIN WADE, DANA LYTLE

DESIGNERS
KEVIN WADE, DANA LYTLE,
ERIK JOHNSON,
TOM JENKINS

CLIENT
MADISON AIDS SUPPORT
NETWORK

Planet Design Company was asked by the Madison AIDS Support Network to design a thank-you gift for the 50 members of its volunteer support group, with an overrun for use as a promotional tool. The studio recommended that a limited-edition poster be produced. To symbolize the group's struggle to overcome AIDS, a theme of "Holding Strong" was selected—supported by a style of illustration and composition that conveyed strength.

By silkscreening two colors on inexpensive tagboard, Planet Design was able to produce 250 posters within its $250 budget.

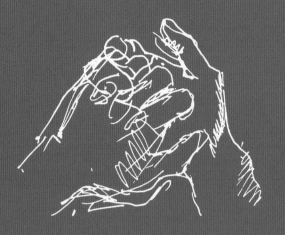

VOLUNTEERS

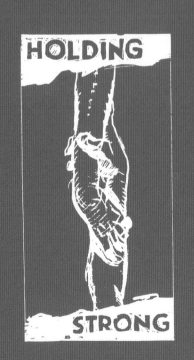

HOLDING
STRONG

HOLD
ING
S+R
ONG

MADISON
AIDS
SUPPORT
NETWORK
VOLUNTEERS
238-6276
DESIGN: PLANET DESIGN CO

MATSUYA — IDENTITY

DESIGN FIRM
PAOS (MOTOO NAKANISHI, CHAIRMAN; YUTAKA SANO, EXECUTIVE DIRECTOR, DESIGN DIVISION)

CO-DESIGNER
MASAYOSHI NAKAJO

CLIENT
MATSUYA CO.

Matsuya, one of Japan's oldest department stores, asked PAOS to help celebrate its 120th birthday in 1989 with a top-to-bottom renovation of the store and the launch of new business and service areas. As part of this project, PAOS refined the graphic identity it had prepared for Matsuya in 1978. The updated design had to maintain the feel and spirit of the 1978 logotype while also injecting a renewed image of freshness. PAOS also had to prepare a variety of applications for the new mark—such as modifications to represent Matsuya's two Tokyo locations separately and together.

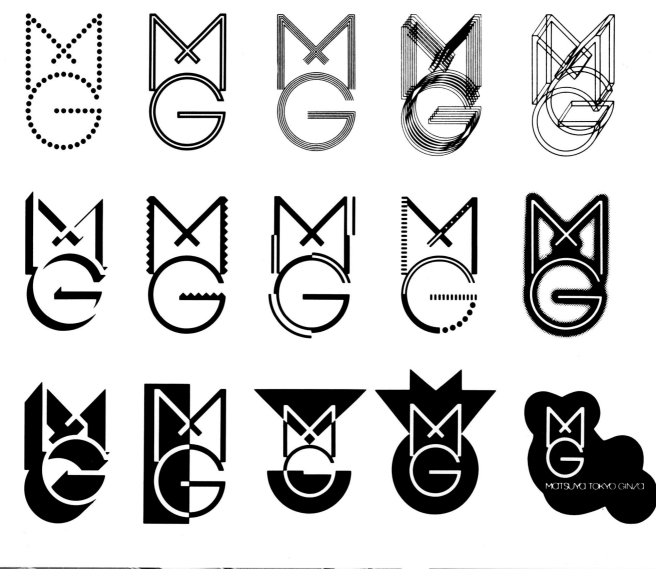

MITSUBISHI ELECTRONICS — IDENTITY

DESIGN FIRM
BRIGHT & ASSOCIATES

ART DIRECTOR
KEITH BRIGHT

DESIGNERS
BILL CORRIDORE,
MARK VERLANDER

CLIENT
MITSUBISHI ELECTRONICS
AMERICA

Mitsubishi Electronics asked Bright & Associates to develop a uniform graphic identity for the company. Its 13 divisions had each been using their own image qualities, type formats, brochure sizes, etc. The Bright consultants examined all of Mitsubishi's existing printed materials. They then developed design concepts for several sizes of collateral materials that would be flexible enough to use both existing and new photography and illustration. Bright & Associates also designed several brochures and a newsletter format. All the new standards were specified in a design system manual.

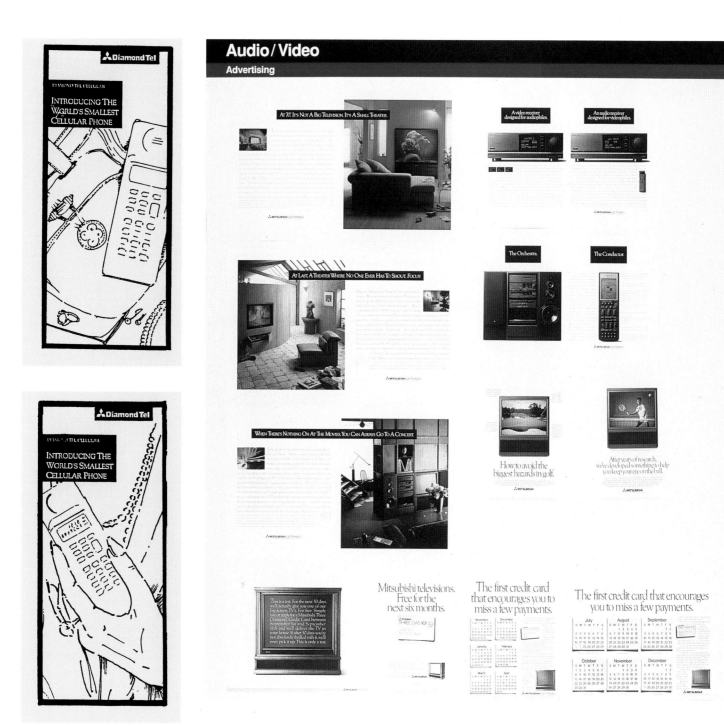

NACHO MAMMA'S — IDENTITY

DESIGN FIRM
SAYLES GRAPHIC DESIGN

ART DIRECTOR / DESIGNER
JOHN SAYLES

CLIENT
NACHO MAMMA'S

When Rich Murillo and his three partners decided to open a Mexican restaurant in Des Moines, Iowa, they contracted with Sayles Graphic Design to develop an identity for their business. Art director John Sayles' first start at a logo proved false: a two-color "Mamma" was rejected because, as Murillo explains, "we wanted people to develop their own idea of who Mamma was and what she looked like." Taking a second stab, Sayles retained some basic elements of the initial design and, in about 12 hours, completed a revised, multicolored logo emphasizing a jalapeño pepper.

Sayles supplemented his new proposal with a model of his vision for the restaurant exterior, including a neon-accented sign, "electric pastel" graphics for the building's canopies, and painted accents on the facade. Encouraged by his clients' ecstatic reaction to the model, Sayles used the same color palette as he developed collateral materials (such as menus) and design ideas for the restaurant's interior (a 15-foot jalapeño piñata hanging from the ceiling, a hand-painted graphic snake covering an entire wall, and corrugated steel wall coverings).

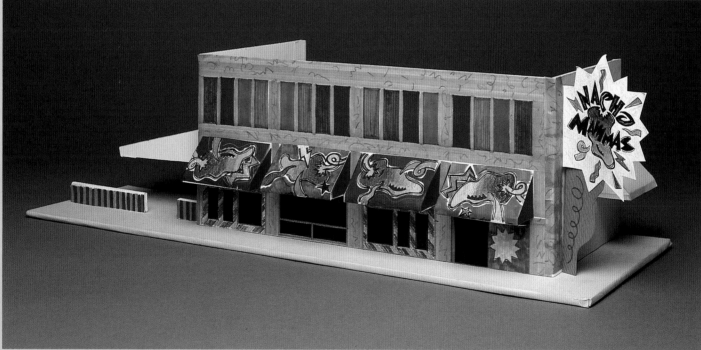

NORTH CAROLINA ALTERNATIVE ENERGY CORPORATION – ANNUAL REPORT

DESIGN FIRM
SALLY JOHNS DESIGN

CLIENT
NORTH CAROLINA
ALTERNATIVE ENERGY
CORPORATION

The North Carolina Alternative Energy Corporation needed an annual report design that would help convey the AEC's activities and objectives in an approachable manner and demonstrate how its state-wide efforts can affect the entire globe.

Sally Johns Design developed a yearbook / clip album approach to illustrate different projects in which the AEC is presently involved. After considering various covers that portrayed specific types of energy usage, the studio and client agreed on the more comprehensive idea of a textured collage effect of a globe.

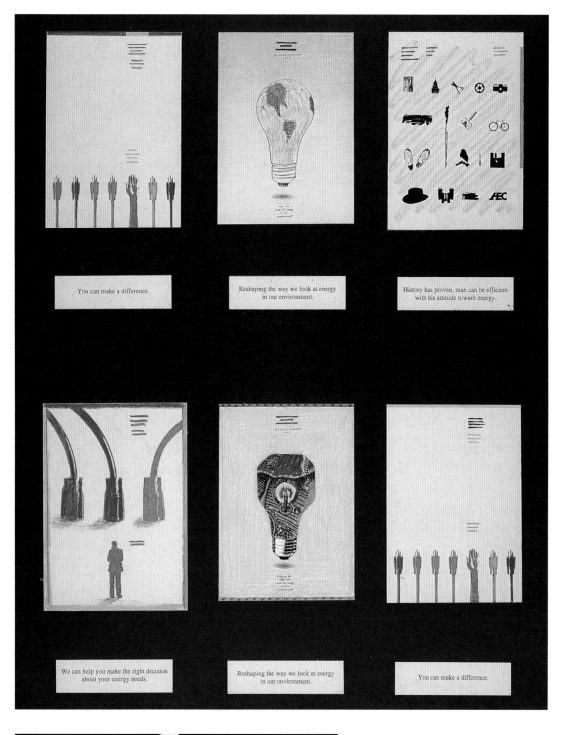

You can make a difference.

Reshaping the way we look at energy in our environment.

History has proven, man can be efficient with his attitude toward energy.

We can help you make the right decision about your energy needs.

Reshaping the way we look at energy in our environment.

You can make a difference.

Once the cover approach was determined, a basic outline of the contents was devised and a typical inside page comp was prepared. High-contrast photocopies, magazine cutouts, drawings, and color photographs were combined over a textured background to form illustrations for each section opener. These same treatments were used elsewhere, in initial caps, in the numerals of the table of contents, and in the descriptive icons of the state map.

These illustration comps were reviewed by the client and then revised as needed. Responding to client feedback, the typographic treatment used throughout the report was kept short, concise, and organized.

The copy you are reading reflects the type face and the point size of the actual copy. The intent of this copy is to give the reader an idea of how the type will look on the page before the actual copy is set. The copy you are reading reflects the type face and the point size of the actual copy. The intent of this copy is to give the reader an idea of how the type will look on the page before the actual copy is set. The copy you are reading reflects the type face and the point size of the actual copy.

Bhe copy you are reading reflects the type face and the point size of the actual copy. The intent of this copy is to give the reader an idea of how the type will look on the page before the actual copy is set. The copy you are reading reflects the type face and the point size of the actual copy. The intent of this copy is to give the point size of

The intent of this copy is to give the reader an idea of how the type will look on the page before the actual copy is set.

Copy. The intent of this copy is to give the reader an idea of how the type will look on the page before the actual copy is set. The copy you are reading reflects the type face and the point size of the actual copy. The intent of this copy is to give the reader an idea of how the type will look on the page The intent of this copy is to give the reader an idea of how the type will look on

A

Ioria Young, a fifth grader at Carrboro Elementary School, generates electricity while her classmate Cranston Farrington and AEC project manager Keith Aldridge look on. Turning on light bulbs with their own muscles gives young people a new way to understand energy. What does energy efficiency mean? It means that much less pedal power is needed to light up a compact fluorescent bulb than a standard incandescent — even though they both give off the same amount of light.

A copy you are reading reflects the type face and the point size of the actual copy. The intent of this copy is to give the reader an idea of how the type will look on the page before the actual copy is set. The copy you are reading reflects the type face and the point size of the actual copy. The intent of this copy is to give the reader an idea of how the type will look on the page before the actual copy is set. The copy you are reading reflects the type face and the point size of the actual copy. The intent of this copy is to give the reader an idea

of how the type will look on the page before the actual copy is set. The copy you are reading reflects the type face and the point size of the actual copy. The intent of this copy is to give the reader an idea of how the type will look on the page before the actual copy is set. The copy you are reading reflects the type face and the point size of the actual copy. The intent of this copy is to give the reader an idea of how the type will look on the page before the actual copy is set. The copy you are reading reflects the type face and the point size of the actual copy. The intent of this copy is to give the reader an idea of how the type will look on the page before the actual copy is set.

and the point size of the actual copy. The intent of this copy is to give the reader an idea of how the type will look on the page before the actual copy is set. The copy you are reading reflects the type face and the point size of the actual copy. The intent of this copy is to give the reader an idea of how the type will look on the page before the actual copy is set. The copy you are reading reflects the type face and the point size of the actual copy. The intent of this copy is to give the reader an idea of how the type will look on the page before the actual copy is set. **ENERGY** eading reflects the type face and the point size of the actual copy. The intent of this copy is to give the reader an idea of how the type will look on the page before the actual copy is set. The copy you are reading reflects the type face and the point size of the actual copy. The intent of this copy is to give the reader an idea of how the type will look on the page before the actual copy is set. The copy you are reading reflects the type face and the the actual copy. The intent of this copy is to give the reader an idea of how the type will look on the page before the actual copy is set face and the point size of the actual copy. The intent of this copy is to give the reader an idea of how the type will look on the page before the actual copy is set. The copy you are reading reflects the type face and the point size of the actual copy. The intent of this copy is to give the reader an idea of how the type will look on the page before the actual copy is set.

Robert K. Koger

NORTH CAROLINA
COMMUNITY FOUNDATION — IDENTITY

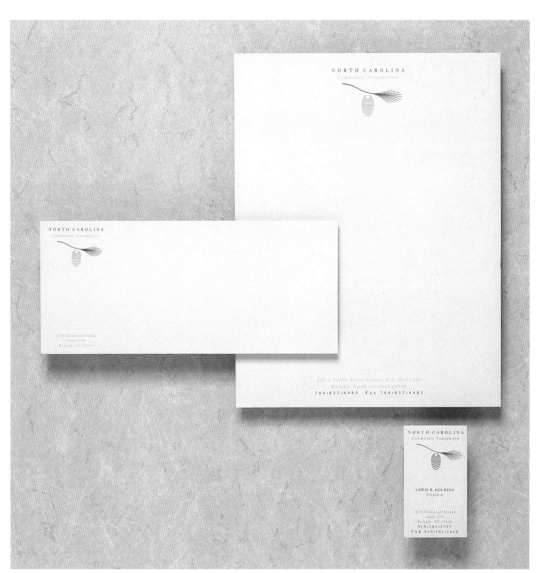

DESIGN FIRM
SALLY JOHNS DESIGN

CLIENT
NORTH CAROLINA
COMMUNITY FOUNDATION

The North Carolina Community Foundation needed a new logo symbolizing its community involvement across North Carolina. Designers Sally Johns and Jeffrey Dale quickly knew the elements they wanted to use to illustrate that commitment: a pine cone, representing the state tree, and interlocking hands, to convey caring and helpfulness. Through a variety of sketches, they developed a sophisticated, yet friendly, graphic that would appeal to a statewide audience. Once the logo was approved, the design studio prepared new stationery and accompanying materials for the foundation.

PAULA ABDUL – ALBUM COVER

DESIGN FIRM
MARGO CHASE DESIGN

ART DIRECTORS
MELANIE NISSEN / VIRGIN
RECORDS; MARGO CHASE

DESIGNERS
MARGO CHASE, NANCY
OGAMI

PHOTOGRAPHER
ROBERT LOBETTA

CLIENT
VIRGIN RECORDS

Designer Margo Chase began work on an album cover for recording artist Paula Abdul by meeting with Abdul and Virgin Records' Melanie Nissen to discuss different approaches for the cover using photographs that had already been shot. With a two-week deadline, Chase presented various packaging ideas for a compact disc, phonograph album, and cassette. One CD tall box slid apart, revealing a tall booklet with a photo of Abdul's eye. A second had gold and red lettering painted on a translucent plastic sleeve. Cover ideas that would be printed more conventionally included a photo of a dancing Abdul and the comp that became the final cover. Besides retouching, and adjusting the logo size, few changes were made to transform the selected comp into the final cover.

RACHELI'S — IDENTITY

DESIGN FIRM
GRAFIK COMMUNICATIONS

CREATIVE DIRECTOR
JUDY F. KIRPICH

ART DIRECTOR
SUSAN ENGLISH

DESIGNER
MICHAEL SHEA

**PHOTOGRAPHER FOR
PRODUCT SHOT**
DAVID SHARPE STUDIO

CLIENT
RACHELI'S

Racheli's commis-
sioned Grafik
Communications to develop
a mark that would suggest
both quality coffee and fast
service—two traits demanded by the professional
clientele that the downtown
Washington, D.C., coffee
house sought. Racheli's
owner requested a design
that was direct, simple, and
upscale. He gave Grafik
sketches and tile samples
from his interior designer
that conveyed the clean,
bright, sophisticated look he
wanted to project.

RACHELI'S

RACHELI'S

RADIO VISION

DESIGN FIRM
MARGO CHASE DESIGN

ART DIRECTOR
MARGO CHASE

CLIENT
RADIO VISION
INTERNATIONAL

Radio Vision distributes live concert videos. It needed a contemporary mark that gave them a hip but business-like image.

Margo Chase presented three rounds of black-and-white logos. The early directions were based on the company's existing logo, which was a face. Chase then tried typographic ideas and, finally, came around to the idea of a "Vision Man."

Chase prepared two final versions of the logo—a simple one with the type and man symbol, and a more complicated shield version. Depending on the specific application, Radio Vision uses either of these logos.

RADIO VISION

RADIO VISION

RADIO VISION

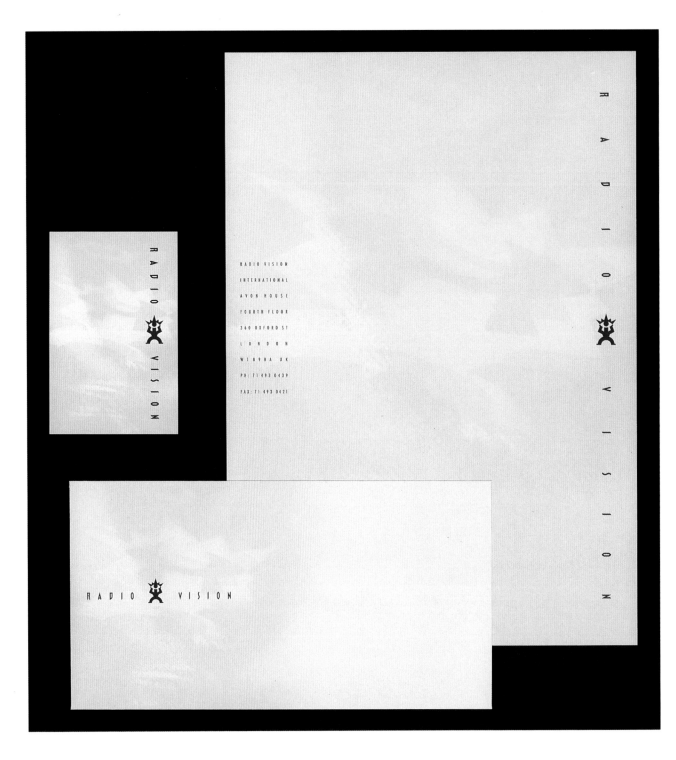

RADIO VISION
INTERNATIONAL
AVON HOUSE
FOURTH FLOOR
360 OXFORD ST
LONDON
W1N 9HA UK
PH: 71 493 0439
FAX: 71 493 0421

Radio Vision chose one of the two approaches to stationery options preferred by Chase. (Her other preferred design featured elongated shields down the sides.) Radio Vision did request one change, however: It liked the green comps that Chase presented but "hated" the eyes that were in the background. In response, the studio created a photogram of Radio Vision's logo for use as background on all pieces of the stationery package.

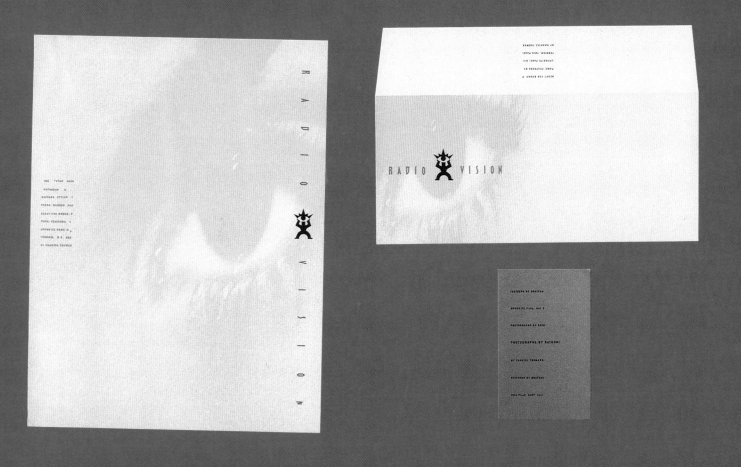

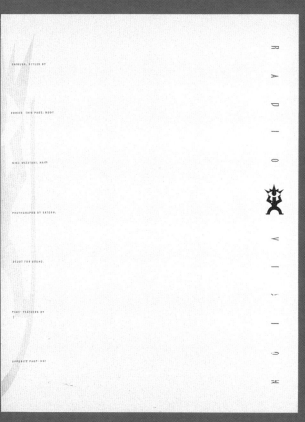

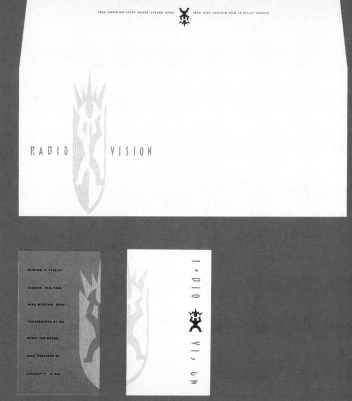

RADIUS — IDENTITY

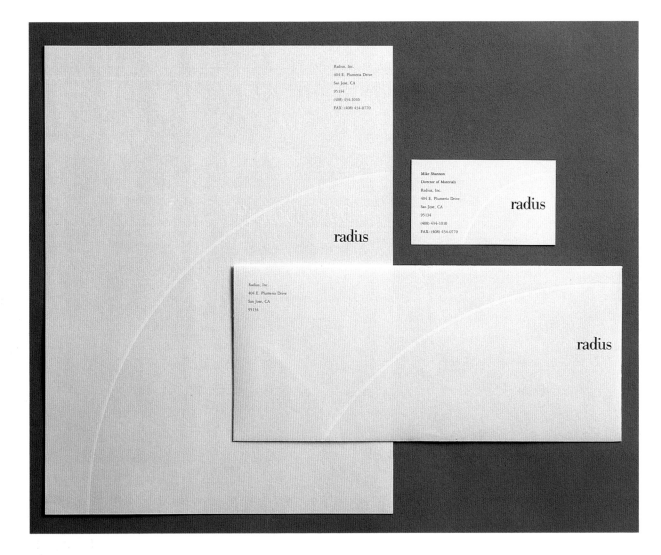

radius

DESIGN FIRM
MORTENSEN DESIGN

ART DIRECTOR
GORDEN MORTENSEN

CLIENT
RADIUS

Radius asked designer Gorden Mortensen to develop a graphic identity for the new computer-hardware company. Believing that it would need to have its official name recognized before the public could identify with any icon, he recommended that Radius adopt as a logo a typographical display of its name. Thus, Mortensen's primary goal was to create a simple yet memorable logotype announcing Radius as a corporation "here to stay." He selected Bodoni as the typeface, believing that it would convey a timeless rather than trendy quality. To add a distinctive touch, he created a ligature of the "i" and "u."

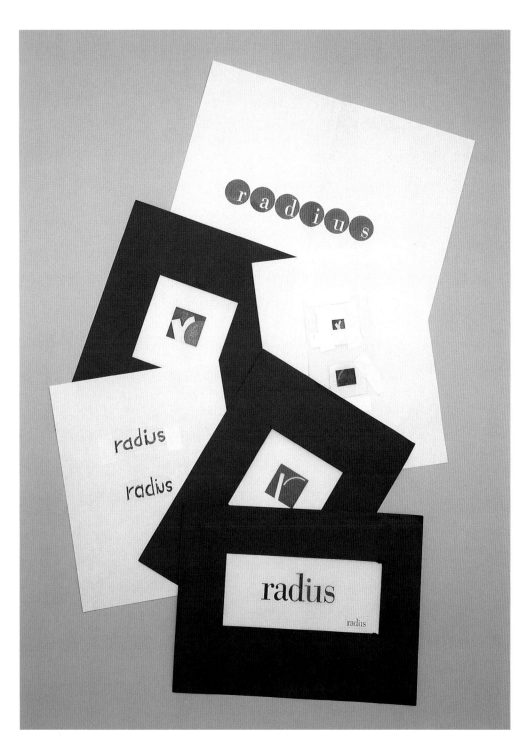

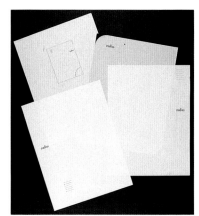

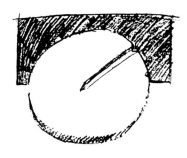

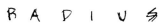

Once the logotype was determined, Mortensen developed a collateral idea of adding a subtle radius / arc across printed materials. This helped give the graphic identity a contemporary look while maintaining an image of solidity and permanence.

In developing a graphic standards manual for Radius to use in implementing its new identity, Mortensen sought a format that would emphasize accessibility for the user. He opted for a relatively small size and a corrugated plastic binder for a clean and contemporary feel.

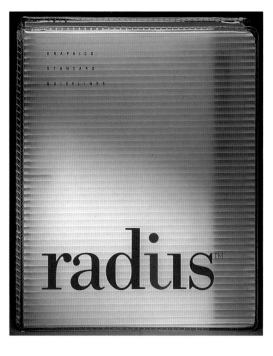

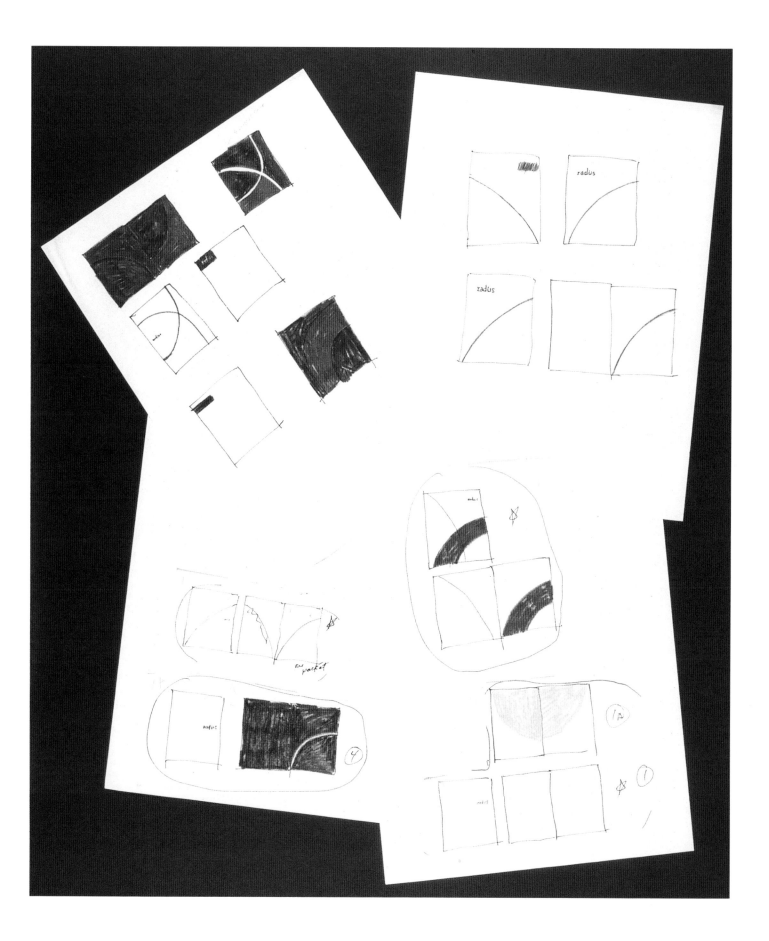

RED TOMATO – IDENTITY

DESIGN FIRM
JOED DESIGN

CREATIVE DIRECTOR
EDWARD REBEK

DESIGNERS
JOANNE REBEK, ED REBEK

WRITER
EDWARD REBEK

ILLUSTRATOR
DAVE VOIGT (JUGGLING /
FLYING LADY MENU
GRAPHIC)

ARCHITECT
KEITH YOUNGQUIST,
AUMILLER YOUNGQUIST

CLIENT
JOE DIVENERE

When Joanne and Edward Rebek were asked to develop a concept for a new Italian restaurant in Chicago, they first generated more than 30 possible names. From that list, restaurant owner Joe DiVenere selected "Red Tomato" because of its connotations of being fresh, ripe, and ready to eat.

The Rebeks then began devising a logo that would capture the owner's sense of fun. Several alternative designs were created, with DiVenere and the Rebeks agreeing on a graphic that featured a casual hand script, varying letter scale, a vertical "RED," and a cartoon Italian plum tomato "O."

RED TOMATO

RED TOMATO

Additional marketing elements also had to convey enjoyment without overusing the logo. Light-hearted visuals were created to suggest "food as fun." The tomato / "O" from the logo was used in large scale for color and impact on the deli's take-out packaging and menu, as well as on the business card and other promotional materials. "Floating tomatoes" and a "juggling / flying lady" were used on items such as T-shirts.

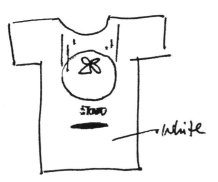

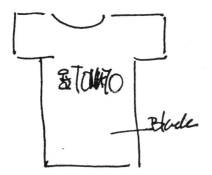

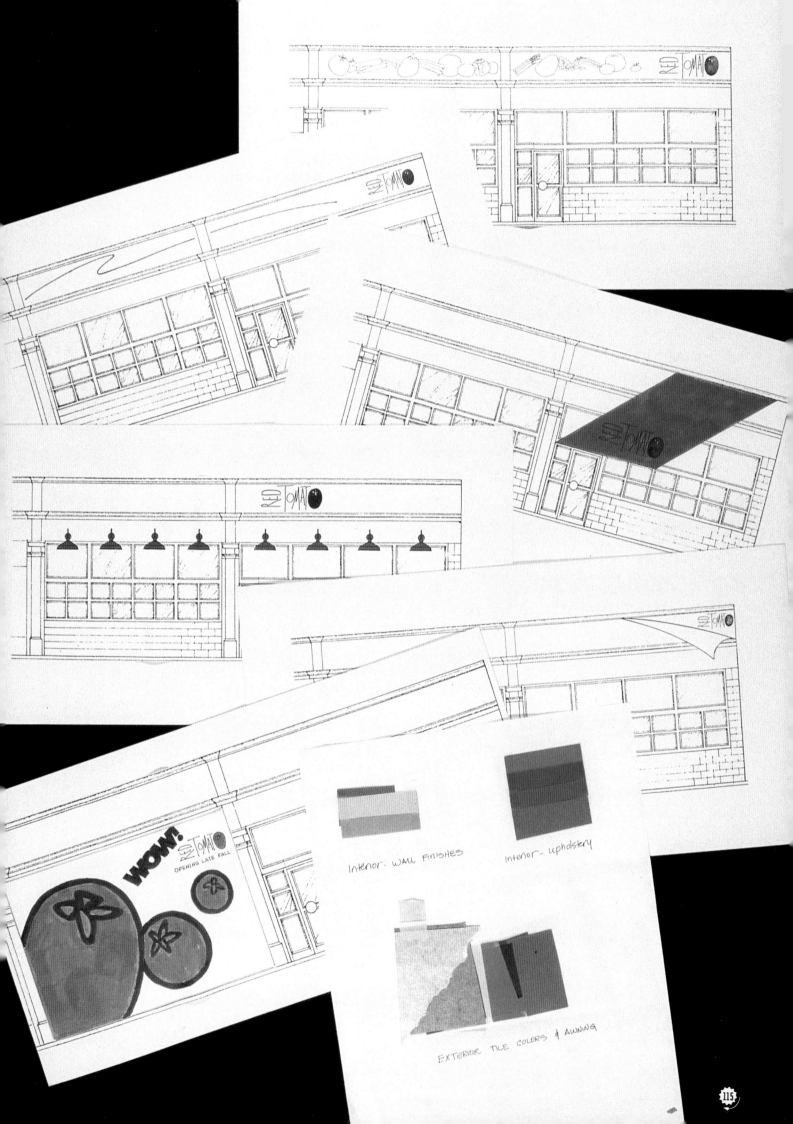

Interior WALL FINISHES

Interior- upholstery

EXTERIOR TILE COLORS & AWNING

WOW!
OPENING LATE FALL

The menu design also needed to convey the "food as fun" concept, but in a more refined manner. For the lunch menu, a statuesque image of a woman was combined with cartoon tomatoes and a green marble-like field. On the dinner menu, the lady reappeared flying through the air with a platter of tomatoes on a rich gold- and black-patterned background.

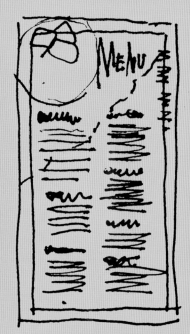

SMACNA — CONFERENCE PACKAGE

DESIGN FIRM
GRAFIK COMMUNICATIONS

CREATIVE DIRECTOR
JUDY F. KIRPICH

ART DIRECTORS
SUSAN ENGLISH,
RICHARD HAMILTON

DESIGNER
RICHARD HAMILTON

**PHOTOGRAPHER FOR
PRODUCT SHOT**
DAVID SHARPE STUDIO

CLIENT
SHEET METAL AND AIR
CONDITIONING
CONTRACTORS' NATIONAL
ASSOCIATION (SMACNA)

When Grafik Communications works each year on a convention package for SMACNA, its designers focus on the history and ethnic background of the convention's host city. They began work on the 1991 package by preparing sketches suggestive of San Antonio and of Texas—in a bold, colorful style appropriate to the Southwest.

Grafik must always produce this package within a fixed budget. For the 1991 materials, it ganged all of the covers together during the printing. The resulting financial savings enabled Grafik to add another color to the project's materials.

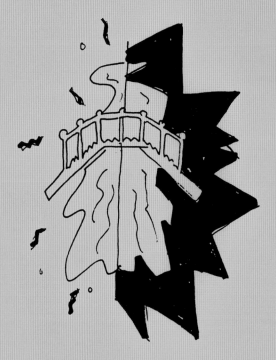

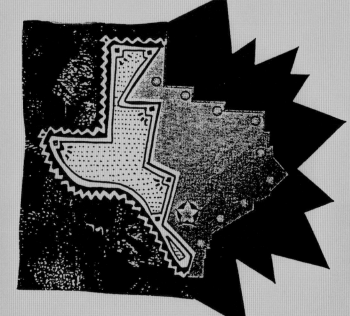

UNLEASH
THE POWER!

Sabc Amnopqr,
Oefahijk 14·17 1991

SEPPELT PARA LIQUEUR – PACKAGING

DESIGN FIRM
BARRIE TUCKER DESIGN
PTY

DESIGNER
BARRIE TUCKER

TYPOGRAPHER
ELIZABETH SCHLOOZ

PRODUCTION MANAGER
SERGIO JELOSCEK

BOTTLE MANUFACTURER
NORD VETRI, ITALY

BOX MANUFACTURER
JOHN INGE

CLIENT
B. SEPPELT & SONS

Australian vintner B. Seppelt & Sons decided to repackage its Para Liqueur to better reflect its heritage as the world's only 100-year-old vintage fortified wine. Designer Barrie Tucker began by designing a bottle in rough form, specifying how to emboss the Seppelt logo, then sent his drawings to a glass manufacturer in Italy for accurate drawing. After a wooden bottle mold and "handmade" glass samples of the bottles were produced and minor alterations were made, Seppelt gave its approval. As a final task, Tucker designed a walnut presentation case.

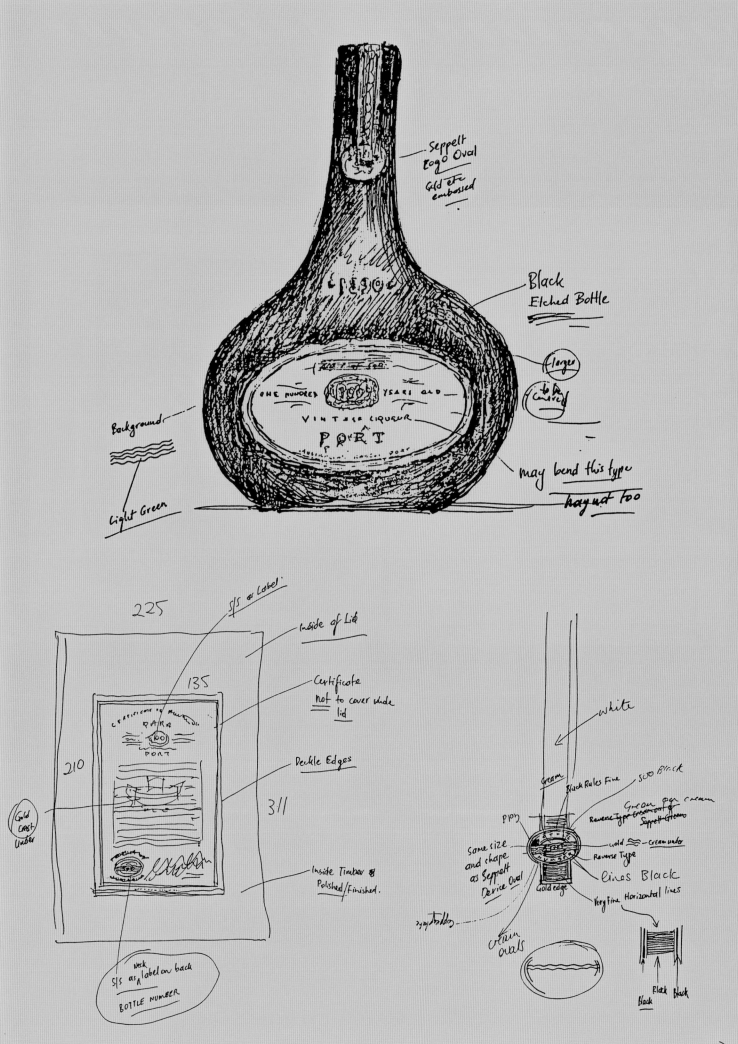

Seppelt
Logo Oval

Gold etc
embossed

Black
Etched Bottle

flange

to be
curved

Background

Light Green

may bend this type

layout too

ONE HUNDRED YEARS OLD

VINTAGE LIQUEUR

PORT

225

S/S as Label

Inside of Lid

135

Certificate
not to cover whole
lid

CERTIFICATE OF MATURITY
PARA
100
PORT

Deckle Edges

210

311

Gold
Crest
Timber

Inside Timber &
Polished/Finished.

S/S as label on back
Neck

BOTTLE NUMBER

white

Cream

Black Rules Fine

500 Black

Green Dev per cream
Reverse Type - Cream over of
Seppelt Green

PPN

Same size
and shape
as Seppelt
Device Oval

Gold ≈ cream under

Reverse Type

Gold edge

lines Black

Very Fine Horizontal lines

Copperplate

cream
ovals

Black   Black
Black

121

SILICON GRAPHICS – IDENTITY

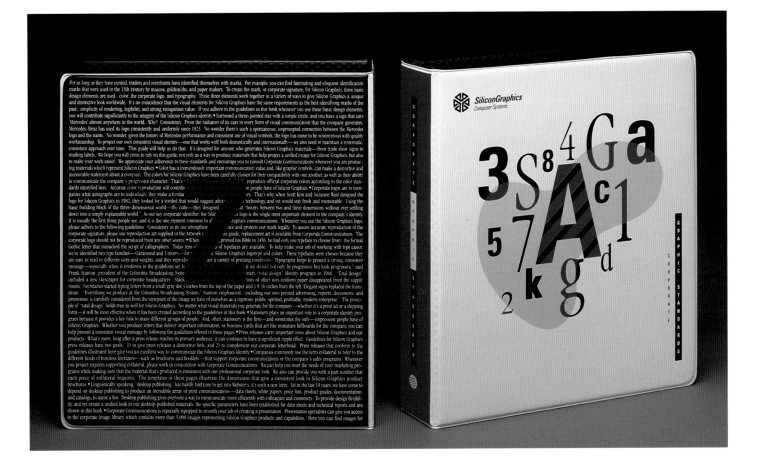

DESIGN FIRM
MORTENSEN DESIGN

ART DIRECTOR
GORDEN MORTENSEN

CLIENT
SILICON GRAPHICS
COMPUTER SYSTEMS

Designer Gordon Mortensen was commissioned by Silicon Graphics Computer Systems to develop a new cover for its graphic standards manual. The binder had to project Silicon's image as a visually oriented company. The binder also had to convey that the standards manual was user friendly and that the guidelines within the manual were not so rigid that the user should feel creatively inhibited. Finally, Mortensen needed to use elements on the binder (type, shapes, colors, etc.) that reflected the subjects covered inside.

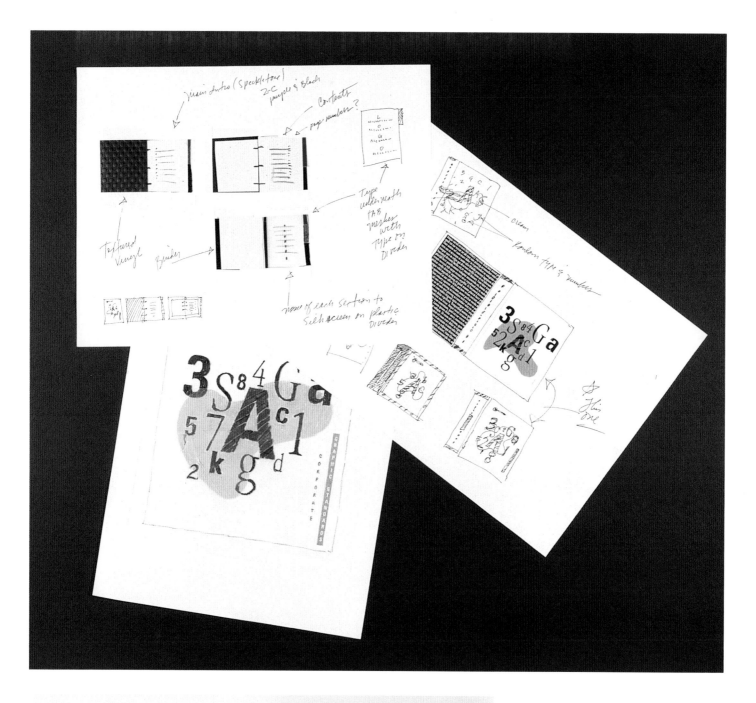

Mortensen was also asked to create a replacement for an existing folder of laser-printed information. Shipped with every computer purchased by customers, it was the first collateral piece that most users saw. Thus, the folder again had to convey Silicon's visually oriented products and image and generate excitement for the new user.

To reinforce the idea that the folder is an important piece designed to be read by the user, Mortensen added Velcro fastening. He gave each insert its own visual treatment to help sustain user interest in all items in the folder.

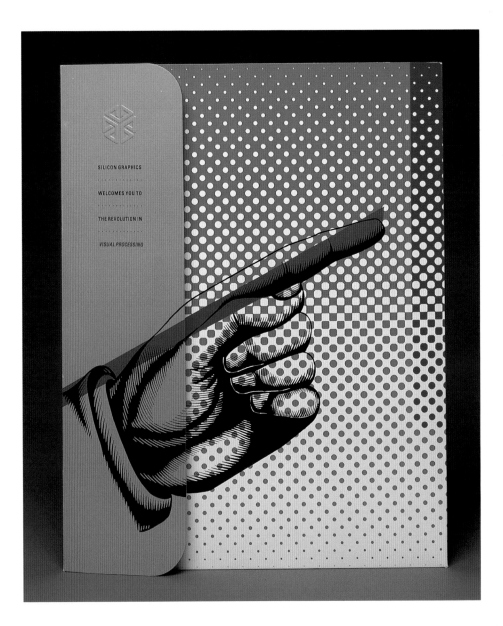

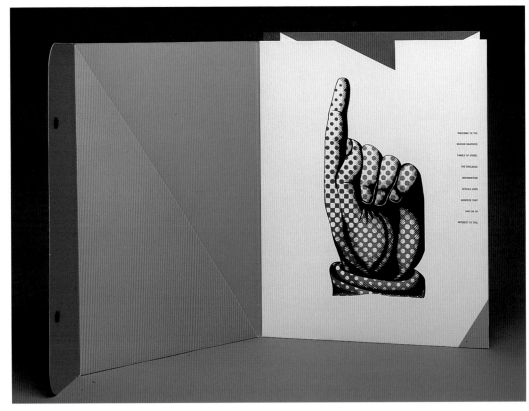

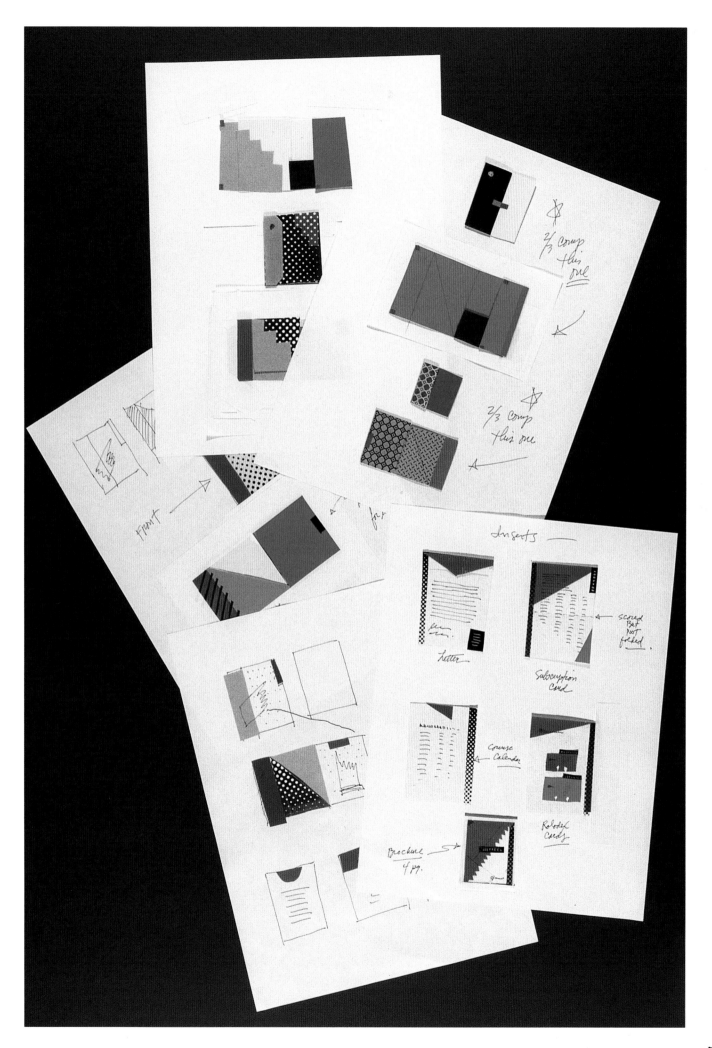

STANFORD CONFERENCE ON DESIGN – POSTER

DESIGN FIRM
MORLA DESIGN

DESIGNER / ILLUSTRATOR
JENNIFER MORLA

WRITER
NANCY THOMPSON

CLIENT
STANFORD ALUMNI
ASSOCIATION

E ach year one designer is selected to produce a combined mailer / poster announcing the Stanford Conference on Design.

When Jennifer Morla was asked to design the announcement for the 1989 conference, she explored a variety of graphic approaches. She decided on— and the client readily agreed to—an approach that depicts bringing men and women of all architectural and design professions together. The leaves, sun, and Hoover Tower symbolize the common images of the Stanford campus, while the trapezoid, pyramid, and circle symbolize basic forms in architecture and design.

Tokyo Broadcasting System — Identity

DESIGN FIRM
SHIMOKOCHI/REEVES

CREATIVE DIRECTORS
MAMORU SHIMOKOCHI,
ANNE REEVES

DESIGNER
MAMORU SHIMOKOCHI

**ADVERTISING AGENCY
ART DIRECTORS**
KUNIAKI MIYASAKA,
MASAHIRO TOBIOKA /
DENTSU, TOKYO

CLIENT
TOKYO BROADCASTING
SYSTEM

The Tokyo Broadcasting System (TBS) wanted a new identity that would suggest diverse programming and appeal to younger viewers. Dentsu, TBS's advertising agency, asked three design studios—in Tokyo, Paris, and Los Angeles—to develop proposals. Of over 200 marks submitted, the agency selected one by Shimokochi/Reeves, a Los Angeles firm.

After sketching numerous, relatively conservative thumbnails, Shimokochi/Reeves decided to try more avante-garde approaches. It prepared—and submitted to Dentsu—three new concepts: "Technology," a "T" containing various graphics; "Vitality," a man holding a lightning bolt; and "Microcosmos," the winning approach, which contained planets and other graphics within the "TBS."

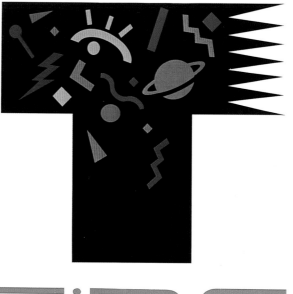

Once the network's
management selected
"Microcosmos" as its new
logo, Shimokochi/Reeves
developed widely varied
applications. These ranged
from stationery and station-
identification broadcast
spots; to signage and the
corporate flag; to vehicle
and satellite usage; and to
such promotional items as
T-shirts, windbreaker jackets,
and credit card-sized radios.

Once the logo was developed, it became clear that the black-and-white version of the new mark was too complex to reproduce well with some usages, so a simplified version without all the interior graphics was created. This mark was used, for example, with the logo for the network's news staff, as signage on the network's headquarters building; and on souvenir golf balls and sport towels.

TONY WILLIAMS — VIDEODISC COVER

DESIGN FIRM
MARGO CHASE DESIGN

ART DIRECTOR
MARGO CHASE

DESIGNER
NANCY OGAMI

PHOTOGRAPHER
MARGO CHASE

CLIENT
BLUE NOTE RECORDS

When jazz artist Tony Williams asked Margo Chase to design a cover for his forthcoming videodisc, he didn't want a photograph of himself on it. This enabled the studio to create a conceptual image reflecting the feel of his music instead.

Chase presented a variety of comps for consideration, including a photogram intended to portray the dark, unpolished look of a New York jazz club and another comp with a transparent blue "TW" logo over a photo of burned trash.

A third comp—selected without alteration—used a double exposure that Chase shot, printed, toned, and hand-colored as the background. Another photo was stripped into calligraphy of Williams' name.

UNIVERSITY OF OTTAWA – CAMPAIGN IDENTITY

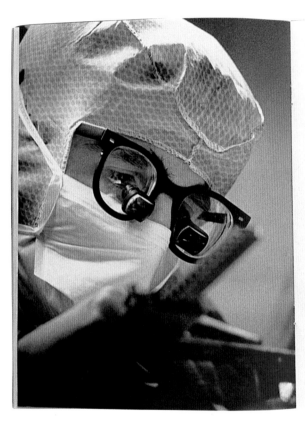

DESIGN FIRM
TURQUOISE DESIGN

ART DIRECTOR
MARK TIMMINGS

DESIGNER
MARIO GODBOUT

PHOTOGRAPHERS
MICHEL CHEVALIER,
PIERRE BERTRAND

CLIENT
UNIVERSITY OF OTTAWA

In 1991 the University of Ottawa launched a Vision Campaign to raise $34 million. University officials asked Turquoise Design to develop a campaign identity reflective of the school's heritage as a premier bilingual university.

Turquoise created a visual story to support the written information in the campaign's publications. Archival images established a traditional tone, while contemporary action shots of students expressed future challenges. The design grid for the brochure was constructed to feature the visual story and allow for innovative placement of English and French text.

Campagne Vision
UNIVERSITÉ D'OTTAWA

Vision Campaign
UNIVERSITY OF OTTAWA

UNE VISION D'AVENIR. L'Université d'Ottawa est la principale université bilingue du Canada. Depuis sa fondation en 1848, les programmes de l'Université et ses plus de 72 000 diplômés et diplômées ont joué un rôle tout particulier dans les rapports entre les populations anglophone et francophone du Canada. ● L'Université est également connue dans le monde entier pour la qualité supérieure de ses programmes d'enseignement et de recherche. Plusieurs chaires, instituts et centres d'excellence figurent parmi les nombreuses réalisations. ● C'est sur cette base solide que l'Université d'Ottawa doit bâtir son avenir. Si elle veut léguer son riche héritage aux jeunes esprits et aux chercheurs doués des générations à venir, l'Université doit agir. ● La Campagne Vision est un programme de financement de 24 millions de dollars qui permettra à l'Université d'Ottawa de poursuivre sa mission unique. ● Donnez généreusement à la Campagne de l'Université d'Ottawa. Grâce à votre participation, l'Université pourra réaliser sa vision d'avenir.

A VISION FOR TOMORROW. The University of Ottawa is Canada's premier bilingual university. ● Founded in 1848, the University's programs and its more than 72,000 graduates have played a special role in building bridges between Canada's English and French cultures. ● The University is also recognized internationally for its superior teaching and research programs. Numerous chairs, institutes, and centres of excellence are among its many marks of distinction. ● It is on these strong foundations that the University of Ottawa must build its future. To ensure that its rich heritage can be preserved for future generations of young minds and talented scholars, the University must act. ● The Vision Campaign is a special $24 million capital fundraising initiative that will help the University of Ottawa to continue pursuing its unique mission. ● Give to the University of Ottawa Capital Campaign. Through your generous support, you can help the University realize its vision of the future.

WALTER HAMADY EXHIBIT – PROMOTION

DESIGN FIRM
PLANET DESIGN COMPANY

ART DIRECTORS
KEVIN WADE, DANA LYTLE

DESIGNER
DANA LYTLE

PHOTOGRAPHER
MIKE REBHOLZ

CLIENT
CHARLES A. WUSTUM
MUSEUM OF FINE ARTS

The Charles A. Wustum Museum of Fine Arts in Racine, Wisconsin, commissioned a monograph to accompany an exhibit of the sculpture, collage, and book art of Walter Hamady. To keep the booklet's design from overshadowing the illustrations of Hamady's work, Planet Design Company used photographs of Hamady's work as the major design element. To stay within their limited budget, Planet quoted this project several ways. In the end, after receiving all the quotes, the designers printed each of the two signatures and the cover of the monograph using different processes.

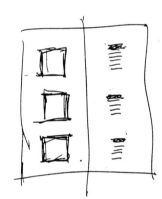

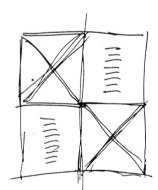

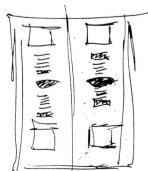

ZEXEL – Identity

·ZEXEL·

Design Firm
PAOS (Motoo Nakanishi, chairman; Yutaka Sano, executive director, design division)

Client
Zexel Corporation

Diesel Kiki, a Japanese leader in diesel systems, asked PAOS to overhaul its identity and select a new name for the corporation. After testing for names that would not be confused with competitors, PAOS proposed ZEXEL—with Z, as the last letter in the alphabet, symbolizing the ultimate, perfection; and EXEL, from the Latin *excellere*, meaning to be better than the rest. Using precision squares to evoke technical perfection, PAOS developed a linear logotype for the new name. Blue (for precision) was designated as the primary color, with red (suggesting advancement) used as an accent.

コーポレートカラー

コーポレートカラー

プレシジョンブルー／アドバンスドレッド

スタンダードカラー

DIC PARTS 1B 2599	DIC 13番 157

Black　White

DIC 13番 65
DIC 13番 123　DIC 13番 547
DIC PARTS 1B 2599　DIC 13番 621 Silver　DIC 13番 620 Gold

プレシジョンブルー＝C-60％×M-40％
アドバンスドレッド＝Y-100％×M-90％

プロモーショナルバージョン

·Z·X·L· ·Z·X·L· ·Z·X·L·

·ZEXEL·
·Z·X·L·
·ZEXEL· ·ZEXEL·
·ZEXEL· ·ZEXEL·
·Z·X·L· ·Z·X·L·

正式社名ロゴタイプ

和文 横組
株式会社ゼクセル

欧文
ZEXEL CORPORATION

和文 縦組
株式会社ゼクセル

関連会社名ロゴタイプ

株式会社ゼクセル販売
ZEXEL SALES COMPANY
株式会社ゼクセル興産
ZEXEL KOSAN COMPANY
株式会社ゼクセルエンタプライズ
ZEXEL ENTERPRISES COMPANY
株式会社ゼクセル物流
ZEXEL LOGISTICS COMPANY
ゼクセル不動産
ZEXEL REAL ESTATE COMPANY
ZEXEL USA CORPORATION
ZEXEL ILLINOIS, INC.
ZEXEL TEXAS, INC.
ZEXEL TECHNOLOGIES USA, INC.
ZEXEL-GLEASON USA, INC.
ZEXEL AUSTRALIA PTY. LTD.
ZEXEL GmbH
ZEXEL-GLEASON EUROPE, S.A.
ZEXEL SINGAPORE CO. PTE. LTD.

最小使用サイズ

株式会社ゼクセル　17mm　　ZEXEL CORPORATION　24mm

株式会社ゼクセル　20mm

·ZEXEL·
·ZEXEL·

·ZEXEL· ·ZEXEL· ·ZEXEL· ·ZEXEL·
·ZEXEL· ·ZEXEL· ·ZEXEL· ·ZEXEL·
·ZEXEL· ·ZEXEL· ·ZEXEL· ·ZEXEL·

·ZEXEL· ·ZEXEL· ·ZEXEL· ·ZEXEL· ·ZEXEL·
·ZEXEL· ·ZEXEL· ·ZEXEL· ·ZEXEL· ·ZEXEL·

·ZEXEL·
ZEXEL
·ZEXEL·
ZEXEL·NOZZLE ·ZEXEL·

·ZEXEL·
·Z·X·L· ·ZEXEL·
·ZEXEL·

·Z·E·X·E·L·
·Z·X·L·
·Z·X·L·

株式会社ゼクセル
ZEXEL CORPORATION

ZEXEL
Corporate Identification
Standards

·ZEXEL·

Studio Biographies

Bright & Associates

Bright & Associates is an international identity and design consulting firm working in the areas of corporate, retail, and brand identity; packaging; graphic communication; and environmental design. The firm was founded in 1977 by Keith Bright, a leading figure in the design community for 30 years.

Bright & Associates has completed successful programs in a broad range of categories, such as financial services, food and beverage, travel and entertainment, sports, and telecommunicatons. It has won numerous design awards from such organizations as the Art Directors Clubs of New York and Los Angeles, the American Institute of Graphic Arts, and the Packaging Designers Council. The firm's work has also been selected for the collections of the Smithsonian Institution and the Library of Congress.

Dale Design

Dale Design was established in 1992 in Raleigh, North Carolina. Though recently formed, this successful graphic design and marketing firm has quickly developed an impressive local and national client base. Owner Jeff Dale is no newcomer to the graphics field. He has ten years' experience in graphic design, advertising, and marketing. He backs his experience with national and regional awards from *Print*, New York Art Directors Club, Washington Art Directors Club, and The American Advertising Federation. Jeff feels client / studio teamwork, research, understanding, hard work and a creative twist are the successful ingredients for quality results. Dale Design plans to continue to broaden its national client base as well as expand into the international market.

Graffito

Graffito, founded in 1985, is a design studio specializing in progressive solutions for corporate, retail, and institutional clients. Based in Baltimore with an office in Philadelphia and a staff of 12, Graffito is a pioneer in the digital design to prepress arena, being the first in the Mid-Atlantic region to incorporate the Lightspeed design system. Clients include SkyTel, MCI, Urban Outfitters, Maryland Institute College of Art, Washington Gas Light, National Public Radio, and Maryland National Bank, among others. The studio has won numerous local, national, and international design awards, and its work has appeared in *Graphis*, *Communication Arts*, *Print*, *AIGA*, *Graphic Design USA*, *How* magazine, *Typography 12*, and Art Director's Club annuals for New York, metropolitan Washington, Philadelphia, and the Advertising Association of Baltimore.

Grafik Communications

Grafik Communications was started in 1977 by Alex Berry, Jr., and Judy Kirpich. It has grown from a small design firm located above a wig shop to a full-service marketing communications group of almost 20 people. The studio has garnered over 100 local, national, and international awards, and its work has been represented in *Communication Arts*, *Graphis*, *Print*, *How*, and *Studio* magazines.

Grafik's broad client base ranges from corporations to museums and other cultural institutions across the United States. The studio's philosophy of design is based on the notion of collaboration. Every project is assigned a creative team. Within that team, ideas are traded, critiqued, and refined. The studio's portfolio of work demonstrates that superior ideas come out of collaboration, not only within the design team but between Grafik's clients, printers, illustrators, and photographers.

JOED Design

JOED Design is a full-service firm and creative resource. Work ranges from corporate identity, print, advertising, and package design. The work has received dozens of awards and is featured in numerous design publications. JOED Design is owned by the husband-and-wife team of Ed and Joanne Rebek, who, after working as design directors at Chicago firms for six years, started JOED Design four years ago.

Margo Chase Design

With an education in biology and medical illustration, Margo Chase never expected to find herself designing logos or album covers for some of the most popular groups of her time. And yet here she is.

Her first job in 1981 designing plain wrap for a grocery store chain was an unlikely beginning. Since then her work has become anything but plain. Her design now is both eclectic and unusual. Her innovative use of photography, typography, and handlettering has served her well in the music industry and has allowed her to expand into such areas as consumer packaging and corporate identity. Her client list includes several major record labels, and she has designed logos and album covers for some of today's best-known popular entertainers.

Michael Doret

Michael Doret, a Cooper Union graduate and native New Yorker transplanted to Los Angeles, has run his own design studio since 1972. His specialty is letterforms and their integration with imagery. Among the works he is most proud of, he lists the six *Time* magazine covers that have made it to the newsstands. Among his favorite works, he lists the covers for Eric Baker and Tyler Blik's *Trademarks of the '20s and '30s* and its sequel *Trademarks of the '40s and '50s*. He is also very proud to have designed the graphic identity for the Graphic Artists Guild. His work has been seen extensively in all the major shows and annuals and features on his work have appeared in many national and international publications and books.

Morla Design

Morla Design was formed in 1984 as a multifaceted design firm offering creative services encompassing print collateral, packaging, identity, logo, signage, and interior architectural design. The studio has been instrumental in the image building of some of the country's largest and most visible corporations.

Morla Design's achievements in print, broadcast, and architectural graphics have been honored in numerous competitions and publications and have received gold and silver medals from the New York and San Francisco Art Directors Clubs. The studio's work is represented in the permanent collections of the San Francisco Museum of Modern Art and the Library of Congress. Most recently, Morla Design received the 1991 Marget Larsen Award for outstanding design achievement.

Mortensen Design

Gordon Mortensen has been an art director and graphic designer for over two decades. He began his career in the field of magazine design. In 1978 he formed his own design firm, Mortensen Design. Over the years the company has distinguished itself in the areas of editorial design, package design, and creation of corporate logos and identities. Mr. Mortensen has won over 300 awards in national and international exhibitions. In 1981,

Communication Arts magazine, one of the best known of the design industry journals, devoted an article to Mr. Mortensen. He has also taught design at San Jose State University and has served as judge at design exhibitions sponsored by the New York Society of Illustrators, American Illustration (New York), and for the Art Annual edition of *Communication Arts* magazine. Mortensen Design includes three designers, and most of its clients are high-technology companies.

NEUMEIER DESIGN TEAM

Marty Neumeier was born in Los Angeles, California, in 1947. He attended Art Center College of Design and later moved to Santa Barbara, where he started the Neumeier Design Team in 1974. In 1984 he moved to his present location in Silicon Valley, just south of San Francisco. He and his team specialize in software packaging and identity for high-technology companies. Neumeier has won more than 200 design awards. Articles on his work have appeared in *Communication Arts, Print,* and the European design journal *Novum Gebrauchsgraphik.* He has also

written numerous articles on graphic design and is a contributing editor for *Communication Arts.*

PAOS

PAOS was founded in Tokyo in 1968 by Motoo Nakanishi, who still leads the company. It has provided consulting services to more than 50 Japanese corporations, each in a different industry. PAOS coined the concept of "metapreneurship" to describe the cluster of competencies and services that make up its consulting practice in helping corporations re-examine the basic assumptions of business organization. PAOS offers services from senior-level management consulting and implementation services to graphic and environmental design.

Today, the PAOS Group has expanded to include PAOS New York, which conducts research for PAOS's Japanese clients in Japan and the United States;

PAOS Boston, an American corporation providing full-scale consulting on the PAOS model; and PAOS Seoul, which manages client projects in Korea.

PENTAGRAM DESIGN

Pentagram is an organization of designers in London, New York, and San Francisco. Pentagram's partners head up their own groups of designers. Each group is autonomous, dedicated to the partner's work. This approach keeps the partner

clearly in front of each job, accessible to clients, accessible to the design team. There are also a number of Pentagram associates. Each is appointed from within and attached to a partner's group, taking direct responsibility with clients for jobs.

The partners subscribe to the organization and its ideals by working for their clients in the name of Pentagram. And the commitment is carried beyond their work for clients to publishing pamphlets and books, advising institutions and governments, sponsoring lectures and educational projects, judging and lecturing.

PLANET DESIGN

Planet Design Company in Madison, Wisconsin, was established in 1989 by designers Dana Lytle and Kevin Wade. In addition to these two partners, the firm's staff consists of two additional designers and a production coordinator. Planet's work principally consists of promotional materials for arts organizations, fashion and business to business clients, annual reports, brochures, posters, corporate identity, logo design, trade shows and signage.

SALLY JOHNS DESIGN

Sally Johns Design is a full-service marketing / communications firm, specializing in evaluating client needs and producing a pleasant and effective finished printed product. Established in 1977, Sally Johns Design has evolved a base of accounts that include complete familiarity and understanding of corporate design, production, and evaluation. It has extensive and varied experience in the design and production of medical, institutional, and commercial marketing and collateral materials. With its six-person staff, Sally Johns Design also has particular depth of service in creative problem solving and in creating and executing direct mail programs, promotional activities and promotional materials.

SAMENWERKENDE ONTWERPERS

Samenwerkende Ontwerpers is a design consultancy involved in graphic and architectural design in its broadest sense. It has offices in Amsterdam and Utrecht, Netherlands. Its client list includes government ministries, cities, companies, museums and foundations.

SAYLES GRAPHIC DESIGN

Des Moines, Iowa-based Sayles Graphic Design was founded in 1985 by John Sayles and Sheree Clark.

They had met a year earlier while Clark was working at Drake University and hired Sayles to work as a freelance graphic designer on a project.

During their first year in business, Clark and Sayles "did quite a bit of charity [pro bono] work," says Sayles—a tradition they have continued. "We had to build a book, and it made sense to use our time creating good work rather than wait for the phone to ring. Now we can contribute our talents to causes we believe in or select the high-profile projects we want to do."

This commitment is reflected by what Clark calls their "Mission Statement": Creativity, Consistency, and Honesty. With a clientele that includes businesses and organizations in Chicago and southern California as well as Des Moines, Sayles and Clark have had plenty of opportunity to test their guiding principles.

Sayles Graphic Design has received numerous honors, including recognition from *Graphis, Print, Communication Arts*, the New York Type Directors, the University and College Designers Association and others.

SHIMOKOCHI/REEVES DESIGN

Mamoru Shimokochi and Anne Reeves have developed identities for many national and international companies. Shimokochi/Reeves takes a strategic approach to the solution of their clients' problems based on developing a real understanding of their clients' products, services, and marketing objectives. This is maximized by the personal involvement of the partners throughout the project. They have earned a reputation for unique and effective solutions to their clients' marketing problems, in many cases resulting in dramatic sales increases.

Shimokochi established his design firm in 1976, with Reeves joining him in 1985. Prior to opening his own office, Shimokochi worked at Saul Bass & Associates for seven years on many of its corporate identity and packaging programs. Reeves, born and educated in England, has had more than 15 years' experience in both design and marketing in London and the United States. Shomokochi's and Reeve's work has received numerous awards for design excellence, and the firm has been published in *IDEA* magazine, *Graphis*, *Print*, and *Visual Identity*.

SIBLEY/PETEET DESIGN

Sibley/Peteet Design, was formed in April 1982 by principals Don Sibley and Rex Peteet. The firm was founded as, and remains today, a creative services studio. The emphasis is on creative and design excellence. The firm has grown from a two-man shop to a staff of 10, currently located in the Dallas Design District. The principals have chosen to keep the firm small to allow personal involvement in all business operations and each individual project. This philosophy has kept the client roster small, but impressive.

Sibley/Peteet Design has received numerous awards over the past nine years for its creativity and design. In addition, the studio's work has been highlighted in several design publications and annuals (e.g., *Communication Arts*, *Print*, *Graphis*, *AIGA*, *New York Art Directors Club*), as well as the permanent collection of the Library of Congress.

SUPON DESIGN GROUP

A four-year-old studio located in Washington, D.C., Supon Design Group is a full-service design firm, specializing in a wide range of graphic arts. Owner and art director Supon Phornirunlit heads the 10-member company and is responsible for overseeing its growth. Its client list is diverse and contains many of the nation's best-known organizations. The studio has been featured in several international publications, including *How's Business Annual*, the American Institute of Graphic Arts' *Journal*, *Designers' Self-Image*, and Asia's *Media Delite*. In the past four years, Supon Design Group has earned over 200 awards from the Art Directors Clubs of both New York and Washington, AIGA, Type Directors Club, American Corporate Identity, DESI, *Print's Design Annual*, and many others. Samples of Supon Design Group's work have been exhibited in England, Germany, Israel, Japan, Thailand, and the United States.

THARP DID IT

Tharp Did It is a five-person design firm on the San Francisco Peninsula. Headed by art director / designer Rick Tharp, the 17-year-old studio specializes in corporate, retail, and restaurant visual identity, and packaging design. Tharp says that having control over all projects and having fun doing it are the rewards of keeping the staff small. The studio's work has been recognized with honors from the New York and San Francisco Art Directors Clubs, the West Coast Show, the American Center for Design and *Communication Arts' Design Annual*. Tharp's posters for BRIO Toys were selected for inclusion in the

Smithsonian Institution's Cooper-Hewitt Museum, and a book he designed for the Swedish toy company is in the permanent collection of the U.S. Library of Congress. The studio has also received a CLIO Award for wine packaging.

BARRIE TUCKER DESIGN PTY

Barrie Tucker, one of Australia's most outstanding and distinguished designers, began his design career in 1966 in Europe, then returned to Australia in the early 1970s and established Barrie Tucker Design. He has since developed a reputation as Australia's premier packaging designer and has won international acclaim for three-dimen-sional imagery and environmental design. In addition to winning numerous national and international awards, he has been elected to member-ship in the Alliance Graphique Internationale, the international association of the world's top graphic designers, illustrators and art directors.

As a result of increased foreign interest in his design, Barrie Tucker has recently expanded his activities internationally by establish-ing offices in Singapore, Tokyo and Cologne, Ger-many.

TURQUOISE DESIGN

Since its establishment in 1980 by executive director Philippe Sigouin and artistic director Mark Timmings, Turquoise Design has earned a privileged place in the Ottawa, Ontario / Hull, Québec marketplace. Its design, production, and marketing specialists work together to produce designs that effectively communicate their desired messages.

Turquoise Design serves clients from both the public and private sectors. It has a particular interest in the humanities and works extensively for institutions promoting arts and culture, international development, education, medical research, and the environment. Reflecting the reality of its base in Canada's national capital region, Turquoise Design provides bilingual service to its clients, taking care to maintain the integrity of both English and French in all aspects of service and product.

246 FIFTH DESIGN ASSOCIATES

246 Fifth Design Associates is a professional graphic design studio that has been servicing a variety of accounts for five years. This diverse client base keeps the daily routine interesting. It deals with retail, small business, large corporations, high-tech industries, government, and associations, all leading to experience in marketing-led design, project planning, coordination, and imple-mentation. Principal and art director Terry Laurenzio and the rest of the studio's personnel are experienced design and business profes-sionals who are able to analyze project objectives and carry them through to the creation of an effective communication piece. Over the past five years, 246 Fifth Design Associates has accumulated a total of 33 local and international awards.

CATEGORY INDEX

Project Index

Studio Index